Biotechnology and the New Agricultural Revolution

AAAS Selected Symposia Series

Biotechnology and the New Agricultural Revolution

Edited by Joseph J. Molnar and Henry Kinnucan

Routledge
Taylor & Francis Group

LONDON AND NEW YORK

First published 1989 by Westview Press

Published 2018 by Routledge
52 Vanderbilt Avenue, New York, NY 10017
2 Park Square, Milton Park, Abingdon, Oxon OX14 4RN

Routledge is an imprint of the Taylor & Francis Group, an informa business

Library of Congress Cataloging-in-Publication Data
Biotechnology and the new agricultural revolution.
  (AAAS selected symposium ; #108)
  Bibliography: p.
  1. Agricultural biotechnology. 2. Agricultural
innovations. I. Molnar, Joseph J. II. Kinnucan,
Henry W. III. Series: AAAS selected symposium ; 108.
S494.5.B563B54 1989   630   88-17376

ISBN 13: 978-0-367-01292-2 (hbk)
ISBN 13: 978-0-367-16279-5 (pbk)

About the Book

The advent of new methods in shaping the performance characteristics of plants, animals, and microbes dramatically expands the possibilities for advances in agriculture -- a new "Green Revolution" in the offing. This book examines the impact of such developments on agricultural institutions, agribusiness, and farmers: What happens when a fundamentally conservative structure is confronted with the possibility of substantially reorganized production systems stemming from radical technological change?

The book is an examination of the way biotechnology is changing the hows and whys of agricultural research -- in university or government laboratories and in corporate research centers. The authors point out that biotechnology is forcing us to redefine not only our perceptions of the origins of life, but also how we view questions of concern to our society as a whole -- questions of safety and environmental impact. In addition, they address the newly important political problems of intellectual property rights and the "ownership" of germplasm, as well as the challenge the new technology poses to the regulatory apparatus overseeing agricultural research and development.

Biotechnology and the New Agricultural Revolution is an important contribution to our understanding of how agricultural research is organized and how knowledge of new technologies is disseminated, how these developments are translated into improvements in farm-level productivity, and the ways in which biotechnology is permanently altering food and fiber production in the world today.

Joseph J. Molnar is professor of rural sociology at Auburn University. He most recently edited *Agricultural Change: Consequences for Southern Farms and Rural Communities* (Westview Press). Henry Kinnucan is assistant professor of agricultural economics at Auburn University. He is the author of numerous articles in *American Journal of Agricultural Economics* and related publications.

Contents

Illustrations

FIGURES

Preface

This volume is based on a symposium at the 1986 Annual Meeting of the American Association for the Advancement of Science. The papers are intended to provide some understanding of the ways biotechnology is reshaping agriculture -- research, regulation, and production -- and the major directions that can be anticipated. The process of technical change in agriculture is fundamental to the advance of human systems, given a growing population and a relatively fixed resource base for food production. Biotechnology represents a quantum leap in potential for world agriculture. It may improve the quality of food and fiber, but changes are altering the way knowledge is generated, the way regulators make decisions about risk to health and the environment, and the choices farmers have for making production decisions. This volume attempts to anticipate the human and institutional adjustments accompanying the expanding capability to alter and manipulate plant and animal life toward agricultural objectives.

Joseph J. Molnar
Henry Kinnucan

Acknowledgments

A number of individuals supported and encouraged this endeavor. David Savold was our patient and helpful editorial liaison at the American Association for the Advancement of Science. His invitation and encouragement played a major role in bringing this volume to fruition. Jean Turney and Angela Reese ably and cheerfully did the word processing for most of the volume. Litchi S. Wu provided valuable editorial assistance and clerical support.

J.H. Yeager, Head of the Auburn University Department of Agricultural Economics and Rural Sociology, has played an enduring role as tolerant and resourceful administrator whose sustained support is evidenced in the production of this volume.

Our work has been underwritten by the Alabama Agricultural Experiment Station and is a contribution to regional project S-198 "Socioeconomic Dimensions of Technological Change, Natural Resource Use, and Agricultural Structure."

1. Introduction: The Biotechnology Revolution

Biotechnology is affecting agriculture in a number of profound ways. New techniques offer different ways of producing existing substances as well as a new array of chemical tools for improving the efficiency of agricultural production processes. It is becoming increasingly clear that the steady increases in productivity that occurred over the past several decades are beginning to level off as the possibilities of present methods are exhausted and biological barriers are encountered (Office of Technology Assessment 1982a). We may be reaching a point of diminishing returns with respect to conventional plant and animal breeding techniques (Thurow 1985).

Biotechnologies present an opportunity to shorten the development time for new varieties and breeds and to expand the potential to install new functions and capabilities (Office of Technology Assessment 1984; Hess et al. 1987). A major objective of biotechnology research in plants is to change the genetic makeup of crops to give them increased resistance to environmental and biological stresses (e.g., heat, drought, nutrient deficiencies, insects, and diseases) so there will be less reliance on the amelioration of these stresses by cultural means (Arntzen 1984; National Academy of Sciences 1982).

The impacts of these developments are being felt both at the farm level (Molnar and Kinnucan 1985) and in the competition between regions for producing various commodities. At the national level, institutional competition for research and development roles in the coming technological revolution is reflected in proposals for revisions in formula funding

1

for the state agricultural experiment stations and for the establishment of a National Institute of Agriculture (Buttel 1985; 1986; Buttel et al. 1984). At the global level, biotechnology is fueling international competition for primacy in the marketing of new technology and the production of food and fiber.

One estimate of the rate of technical change in agriculture is that since 1930, productivity has doubled every 30.4 years in corn yields. The rate of change may be compared to computers in which technology has doubled productivity every 1.1 years, or to jet aircraft technology in which technology has doubled productivity every 13 years (Lienhard 1985). Biotechnology offers the prospect of accelerating the rate of productivity growth in agriculture. Although poor food distribution remains the root of current hunger problems, such increases are vital in the face of exponential world population growth and the need for a corresponding level of growth in the world food supply (Shemilt 1982).

The purpose of this prologue is to examine the nature and sources of biotechnology's social impacts on agriculture. By definition this is a somewhat speculative effort, as the outcomes of biotechnology research are only beginning to be available for use at the farm level. But the products are coming and they will affect, in a number of predictable as well as heretofore unexperienced ways, the economies, industrial organization, and institutional support structures serving agriculture. We can rely on existing theories and empirical research on technical change in agriculture to explain much that will happen on the farm level (Cochrane 1979). But we also must examine institutional shifts that are accompanying a regional, national, and global restructuring of agricultural research and input supply in farming.

ISSUES

Industrial Reorganization

Biotechnology is both altering the structure of existing industries and creating new industries. Perhaps the most dramatic example of restructuring is the seed industry, particularly the production of hybrid corn (Kloppenburg 1988).

Seed from open-pollinated varieties of corn can be saved for replanting, but hybrid seed tends to revert to the character of the original inbred lines after the initial expression of optimum traits in the hybrid generation. The hybrid is proprietary in character as the particular inbred lines can be kept secret (Williams 1986). The farmer is motivated to use the proprietary hybrid over purebred varieties because of the relative advantages of the hybrid in terms of such qualities as superior yield and increased disease resistance. Kloppenburg (1988) argues that these characteristics provide seed companies with the opportunity to break down the autonomy of the farmer, retain the gains to proprietary research, and enlarge profit margins.

Since the 1930s, the rapid adoption of hybrid seed corn has fueled the development of the seed industry. Hybrid sales rose on the same trajectory as corn yields to around the current level of one billion dollars. Because the offspring of hybrid seed are not reliable, farmers were guaranteed to be repeat customers and profits flowed steadily to the seed industry. The sustained profitability of seed companies has led to multinational petrochemical and pharmaceutical corporations. Only Pioneer-Hybrid remains an independent firm.

The experience of hybrid corn has set a precedent and suggests a strategy for private sector control of the seed business. Some crops are dominated by hybrids while others are not. Hanway (1978) estimates that 85 percent of the corn varieties used by farmers in 1977 were privately developed, compared to only 3 percent of the peanut varieties. About 89 percent of soybean varieties were publicly developed. Only 5 percent of the corn seed market is supplied by farmers for home use or from local sale; but more that 60 percent of soybean, wheat, and oats are from these independent sources. Other untoward effects of this shift relate to overdependence on a few varieties that may increase the vulnerability of the nation's corn crop to disease through overly focused genetic backgrounds. Some feel that the Plant Variety Protection Act is one step toward ensuring farmer dependence on external sources of seed supply because it institutionalizes breeder' marketplace rights (Williams 1986). As a result of a recent decision by an appeals board of the U.S. Patent

and Trademark Office, genetically engineering plants,
seeds, and tissue cultures can now be patented.
Previously, plant breeders could obtain protection for
single varieties even though the modification could be
installed in many others. Now a patent can be issued
generically to all varieties with the change (Jaffe
1987).

Industrial concentration in corn and the other
seed industries is likely to be exacerbated by unfold-
ing biotechnical advances. The pharmaceutical and
chemical companies possessing the biochemistry research
base have acquired the seed companies to deliver the
ultimate product of their innovations: the seed. The
programming of the genetic code in new varieties will
proceed on the basis of profitability, but it is not
clear how concerns over environmental sustainability,
genetic vulnerability, and distributive justice will
factor into these equations.

Concentration in the industry also raises the
specter of monopoly control of a nation's seed supply.
Monopoly control is a concern for two reasons: the
potential for predatory pricing to farmers and ecological vulnerability stemming from overdependence on a
narrow set of plant varieties. The discovery of a particularly effective hybrid in a specific crop could
lead to widespread adoption. This may threaten national or world supplies if genetic lacunae later contribute of crop failure. These industries are exerting
clear pressure on the public research sector to focus
on "upstream" grandparent line development and to leave
the "downstream" hybrid variety work to the private
sector (Evenson 1984).

Economies of scale in process technologies provide
additional stimulus for industry concentration. For
example, the average cost of producing bovine somatotropin, a growth hormone that stimulates milk production in dairy cattle, is estimated at $4.23 per gram
for a plant producing .5 million doses per day. Increasing the scale of the production facility to seven
million units cuts estimated production costs by more
than half to $1.97 per gram (Kalter et al. 1984, p.
29). Since the output of a seven million unit facility
is sufficient to treat about two-thirds of the national
dairy herd, the economic incentive to produce BST under
monopoly conditions is clear. To the extent that manufacture of other agricultural biotechnologies requires
large minimum plant sizes to be efficient, economic

pressures to organize industry around a single firm will permeate the agricultural input supply sector. Although there are exceptions, the complementary relationship of human-related biotechnology research to animal-related research favors the earlier development of products for animals instead of for plants. Basic cell processes are better understood for animals than for plants, and plant characteristics have much more complex genetic mechanisms. Private firms already involved in pharmaceutical research can easily move into animal agricultural biotechnologies. Dairy, beef, swine, and poultry will be the leading markets for biotechnology products in the near future. Although the dominant scientific techniques used in animal biotechnologies do not involve release of genetically-engineered materials into the environment, health and food safety issues remain as potential hurdles for both animal and plant products (Brill 1985).

Whether biochemical innovations will significantly accelerate the trend toward fewer and larger farms depends fundamentally on how these technologies affect the cost structure of the typical farm firm. If the technologies require substantial capital outlays so that fixed costs of production rise appreciably, smaller farms will be placed at a competitive disadvantage relative to larger, better capitalized firms because per unit production costs after adoption of the technology is higher for the small firm. Conversely, if the new technology is relatively inexpensive, affecting only variable costs, bias in favor of larger firms generally will not exist. For example BST, with an estimated daily treatment cost of $.25 per crop (Kalter et al. 1984), is a relatively scale-neutral biotechnology, as even the smallest dairy farm can afford to adopt the technology. However, even low cost technologies contribute to industry concentration to the extent that they contain a fixed cost element emanating from the need to acquire and assimilate information about the technology prior to adoption.

Safety Concerns

One fear associated with agricultural biotechnology is the potential virulence of altered organisms or the ability of new organisms to gain a selective advantage in the environment. For decades nonindigenous organisms have been introduced into the United States that contributed greatly to our food sources and provided new ornamental species. Agricultural scientists have been able to create new gene combinations in single organisms -- even new species -- through mutagenesis, cross-hybridization, and other breeding techniques. The argument for confidence in biotechnology safety is that these products are not fundamentally different from products obtained from conventional technology (Brill 1985). On the other hand, we have never had the accelerated power to perform these and other manipulations, nor has the enhanced capability been so widespread around the world.

Two major perspectives seem to be emerging. One argues that biotechnology products present marginal or incremental improvements to organisms that we already have no trouble living with, that the survivability of altered cells in the environment is very low, and that few if any practices beyond those already implemented are necessary. Alexander (1985) maintains that the probability of a deleterious effect from a genetically engineered organism is a product of six factors: release, survival, multiplication, dissemination, transfer, and harm. The odds of simultaneous occurrence of all these conditions is low. The concern here is that regulatory overreaction and inefficiencies will stifle innovations and hurt the competitiveness of U.S. industry.

Large-scale release biotechnologies seem to be the most worrisome and controversial (Khachatourians 1986). These involve general release of a genetically engineered microorganism into the environment and thus have a high potential for imposing costs on others. Some bacteria that colonize crops, for example, serve as a nucleus for the formation of ice crystals thus making crops more vulnerable to low temperature. A proposed experiment to introduce bacteria stripped of the ice nucleus forming gene into farm fields was stopped by a court ruling that required a full environmental impact statement. The opposition to the ice-minus bacteria experiment argues that we simply know too little about the consequences of allowing novel organisms to be introduced willy-nilly into the environment. Standard testing protocols have not yet been developed, which makes it difficult to assess the risks of large-scale release technologies (Office of

Technology Assessment 1982b).

Most scenarios mentioned by scientists as examples of the potential hazard from biotechnology applications depict accidents resulting from unintended incursions into the ecological system (Perrow 1984). Biotechnology may create interactions between systems not previously linked, which perhaps could not be foreseen to be linked. Perrow cites the unexpected effects of DDT and related pesticides as a precedent for concern. The danger that Rachel Carson (1962) publicized was not direct poisoning, which is observable and well-understood, but the magnification of the substances in living tissues as they move up the food chain. Thus, the catastrophic or harmful potential of biotechnology lies in unintended interventions into the ecosystem.

Experiments show that rDNA research could create a vehicle for the transmission of hazardous traits (Perrow 1984). Although of low probability, the idea is that the peculiar and subtle complexities of recombinant organisms might lead to serious health hazards simply by interacting with biological systems in new unanticipated ways. The pressures of economic competition for patents and commercial applications would seem to exacerbate these risks (Kenney 1986).

Farmers and Biotechnology

Biotechnology products are only beginning to become available at the farm level (Prentis 1984). Most of these products do not involve application of advanced technology in and of itself. Instead they represent substances improved or made more economical by biotechnology advances used in the manufacturing process. Silage inoculants, microbial treatments for plant protection and soil improvement, animal vaccines, growth hormones, and other pharmaceuticals are examples of items previously available but that are now improved or more widely accessible through biotechnology (Molnar, Kinnucan, and Hatch 1986).

A number of microbial agents useful for crop production have been available and in use for many years (Jeffries 1987). Farmers in some commodities have extensive experience with organisms as production inputs, yet the environmental impact of new organisms is a central issue (Kvistigaard and Olsen 1986).

Health and safety concerns about these products relate more to the unknown consequences of genetically-engineered augmentations to the organisms (Alexander 1985; Perrow 1984) and not generally to the impacts on farmers or farm industries. All biotechnology innovations tend to be supply-increasing, although some potentially dramatic improvements in productivity have been seized upon as a threat to particular categories of farmers and certain industries. Nevertheless, the implications of new technology can be quite different for each individual farm, given the source and nature of the microbial, pharmaceutical, or chemical tool, as well as the economic, human capital, and life cycle situation of the farm operator.

Six major avenues of change affect farmers: biotechnology innovations in process technology that lower manufacturing costs for farm inputs; qualitative improvements to existing farm inputs; new classes of input innovations; up-stream biotechnology developments that alter the demand for farm products; concentration of control among a few multinational seed corporations; and displacement of Third World farmers when biotechnology creates domestic substitutes for their crops.

Cost-reducing. The happiest biotechnology consequences for farmers stem from biotechnology innovations that lower the cost for vaccines, chemicals, or other inputs already in use. In this case, biotechnology allows an old product to be manufactured in a more economical way. Suppliers will not pass all savings to the farmer, but competition among suppliers generally benefits the producer and ultimately lowers production costs.

Biological innovations tend to be associated with increases in output per land area; they are generally land-saving, not labor-saving. Biotechnology products will be favored where land prices are relatively high. In an international context, agricultural producers in land-short Japan, Taiwan, and perhaps some European countries will be most inclined to implement these technologies.

Performance enhancement. Biotechnology may enhance the performance of existing inputs by improving specificity, shelf life, or other utility dimensions that benefit farmers. But farmers often pay premium prices for marginal gains, at least in the beginning (Molnar and Kinnucan 1985). No basic change in the root mechanism of the product is involved, but biotechnology increases its efficacy or ease of use. Improvements in the quality and performance of pharmaceuticals and microbial agents offer clear benefits for farm industries and early-adopting farmers.

New Inputs. New classes of input innovations stemming from biotechnology present a mixed picture for the agricultural producer. These products involve a discrete shift in the production process and may require the implementation of a qualitatively different technology package. Bovine growth hormone, nitrogen-fixing soil inoculents, or frost-inhibiting bacterial plant treatments provide examples.

Inequalities are associated with time and spatial lags in the availability and adoption of any innovation. Early-bird farmers capture so-called adoption rents (Cochrane 1979). Some farmers are in locales better served by markets and infrastructure supporting the use of the innovation (Brown 1981). Farmers in marginal production areas for a particular commodity may have less contact with other producers experienced with the innovation and therefore receive no peer-group support for its use. In turn, suppliers may delay making a product available in some areas where few producers exist as it constitutes a "thin market" not justifying marketing costs. Local suppliers may exhibit a similar reluctance to stock or promote a new product. Furthermore, products may have certain agro-climatic specificities that create regional disparities in effectiveness. Thus, differences in timing and circumstances can widen existing inequalities among farmers.

New classes of inputs often require increased human as well as monetary capital for successful implementation (Rogers 1982). Where complex situational assessments and technical expertise are required to obtain full benefits of an innovation, less-educated and less-wealthy farmers face additional barriers to participation in the benefits of new technology (Miller 1983). Better-capitalized farmers can hire expertise if they do not already possess it. This category of biotechnology innovation is most apt to widen differences between large and small farms.

Biotechnology innovations tend to be supply-increasing innovations. Following Cochrane's (1979) analogy of the agricultural treadmill, late adopters face higher production costs and lower product prices. Early-bird farmers have already cut their costs, enjoyed wider profits in their period of initial adoption, and are ready to respond to the next wave of technical innovation.

The supply-increasing consequences of an innovation are particularly dramatic in the case of bovine somatotropin (BST). If approved and marketed, the growth hormone promises a quantum jump in milk production and a dilemma for both the dairy farmer and the taxpayer (Fallert et al. 1987). Non-adopters will see their adopting neighbors enhance production by at least 20 percent. Dairy farmers, if projections are accurate, will have strong economic incentives to implement the technology in their operations. The taxpayer is already supporting milk prices; BST could increase the cost payments to farmers by $90 million in 1996 if current support levels were maintained (Fallert et al. 1987).

If government payment levels were reduced, the industry would likely experience a dramatic shrinkage in farm numbers. Fewer and larger dairy farms would produce either the same or a greater amount of milk. Supply-increasing biotechnology innovations tend to move agricultural industries toward more concentrated, industrialized structures, and biotechnology is no exception. Although many factors are at work, biotechnology innovations also may accelerate trends toward vertical integration in the food system, where supermarkets operate their own dairy processing plants and possibly their own farms.

Concentration of production and the economies of large-scale facilities also tend to shift production areas. Although the regional economics of BST are not fully apparent, Wisconsin and other states with many family-scale dairy farms have good reason to be concerned about BST and increased competition from Florida, California, and other industrialized dairy production areas.

Substitution effects. Farmers are affected by biotechnology innovations that shift demand among commodities through substitution effects. Currently, soft drink and other food manufacturers are replacing sugar with high fructose corn syrup (Crott 1986). The world sugar market is highly depressed, the taxpayer is making large payments to U.S. sugar producers, and many Third World nations are scrambling for alternative export crops. Similarly, advances in enzyme technology may allow greater substitutability among vegetable oils, affecting palm, peanut, and other oil markets (Tonalski and Rothman 1986). In both cases, advances in food substitution biotechnology create competition among heretofore nonequivalent agricultural commodities and bring downward pressure on prices. Dairy farmers could see competition as biotechnology processes come to allow the economical substitution of vegetable protein for milk protein. From the crop farmer's perspective, biotechnology advances alter the economics of using particular crops so they could serve as fuel, chemical feedstock, or food items creating new markets and production opportunities (van den Doel and Junne 1986).

Agribusiness concentration. Many factors serve to create large seed companies and maintain their position once it has been established. The research and development costs of biotechnology research may have increased the minimum operating scale of seed companies --the level at which production costs are minimized (Rees 1985, pp. 109-110). Economies of integration encourage companies to eliminate middlemen by vertically linking their operations. The production and distribution process can then be better coordinated to markets, and profits can be directed toward favorable tax and regulatory environments.

The sheer size of the investment in large-scale biotechnology is a major entry barrier for new firms in the industry. Established firms also tend to have their position reinforced by control over technical knowledge and basic scientific discoveries. In a field where technological change is rapid, large companies have a clear economic advantage and can afford to support research and developemnt programs to maintain their technological superiority. Such monopolies can be sustained through the patent laws (Rees 1985, p. 110). Biotechnology initially attracted a great deal of venture capital to small firms but the uncertainties of product development and regulatory delay favor the large firm in the long run.

Third World farmers. Another set of concerns about biotechnology and the farmer reflect the needs of gene-poor industrial countries and gene-rich nations of the Third World (Kloppenburg and Kleinman 1987, pp. 1-9). Biotechnology offers a high potential for improving the world's food supply. It also presents the possibility of expanded corporate control over Third World economies and a diminished role for some food-producing countries in the world market place (Yoxen 1984; Doyle 1985).

Hybrid plant varieties shift marketing control upstream to the seed companies and away from the farmer. Linkages between plant varieties and certain proprietary inputs may constrain farmers to technology packages that increase farmer dependence on sole-source input suppliers. Although Third World farmers may produce more food with biotechnology products, part of the profits will be shipped to the developed countries -- the home of multi-national corporations. At issue is the relative share received by the farmers, the developing countries, and the multinational corporations.

A major factor is the source of the raw material for developing high-performance plants. Farmers in developing countries pay more for superior varieties from multinational seed companies that paid little or nothing for the genetic stock used to develop hybrids or patented varieties. Access to, control over, and preservation of plant genetic resources have emerged as the subject of international debate and conflict (Doyle 1985).

Biotechnology corporations search the globe for plants that possess desirable properties to be implanted in commercial food crops (Yoxen 1984). Seed companies take common heritage resources, use biotechnology to convert them into marketable private property, and sell it back to the Third World donors of germ plasm resources. Developing nations resent repurchasing their value-added resources from multinational corporations. Nevertheless, most exchanges are multilateral and the less developed nations do have access to a broader pool of genetic resources than they would likely have if acting on their own behalf. Recognition of a nation's sovereignty over its indigenous plant resources, and creation of a system of compensation might be one strategy for redressing a significant inequity between developed and developing nations (Kloppenburg and Kleinman 1987).

The search for a just solution that equitably compensates government resources, intellectual capital, and the stockholder, while supporting the continued development of improved seed reserves is a difficult task. Somehow the need for better yields from a fixed land resource to feed a growing population must be reconciled with ecological diversity and the need of

Third World farmers to minimize risk rather than to maximize production.

OVERVIEW OF THE VOLUME

The twelve papers in this volume represent a unique aggregation of authors from diverse institutional settings who have common threads of insight into the way technical change is driving institutional, social, and economic change in agriculture. Together they represent a timely, comprehensive, and interdisciplinary perspective on the rapidly accelerating changes affecting food production systems in the world today.

Part One focuses on the institutional aspects of the biotechnology tide. Agriculture in the United States is served by an apparatus of agencies and universities unparalleled in any other sector of the economy. Connections between universities and private firms have grown in number and complexity as patent laws and galloping technology have held out the prospect of rapid economic gain. Busch and Lacy profile the shifting of industry-university relations, raising some disturbing questions about who will be left behind as economic and knowledge bases increase inequality among institutions, firms, and farmers.

The new potentials and unknown safety threats of biotechnology products have rapidly altered the regulatory apparatus that approves new items. Decision processes are still not firmly established for new items based on genetically engineered microbes, plants, or animals. Hatch and Kuchler review the historical context of health and safety regulation in agriculture, as well as the dilemmas currently facing the regulatory, industrial, and research communities.

Major concerns stem from the apparent speed at which new products come on the market and the unknown basis for assessing the risk associated with new biotechnology-based production methods for old products. Kenney outlines a theory of regulation in terms of tension between private and public interests. Industries are closest to the regulatory process and are the first to gain or lose by a decision, while the public tends to be underinformed about what is at stake.

A second set of chapters examines the consequences of biotechnology for farmers. Particular attention is paid to an imminent biotechnology product: Bovine somatotropin (BST), which promises to exert dramatic effects on the dairy industry.

Kinnucan and his collaborators address the general issue of technical change in agriculture and the ways technical change alters the relative position of farm firms and the structure of industries. Farmers gain in fits and starts, and other participants in the food systems usually capture a disproportionate share of the benefits of technical advance. The basic framework identified here has repeated application to the specific issues addressed throughout the volume.

Knutson and his colleagues examine the dairy industry as one of the most technically responsive sectors in agriculture. BST and other changes promise to extend far-reaching adjustments in the number, size, and location of dairy farms.

Complementing an economic examination of bovine growth hormone, Buttel and Geisler detail the controversial social effects of this specific biotechnology innovation. They stress the importance of _ex ante_ assessment of emerging technologies, particulary in terms of yet-to-be substantiated claims about scale neutrality. They are particularly concerned about shifts in the international division of labor in agriculture associated with rapid and differential adoption of new technologies.

Terrill explores connections between improvements in animal agriculture and increases in farm income. He is particularly optimistic about the prospects for small ruminants to improve food systems in this country and abroad.

The competitiveness of U.S. agriculture is a major concern for a number of reasons. Agriculture has been a major credit in the balance of trade and a source of strength for the dollar. Small ebbs and flows in agricultural exports have a great influence on U.S. commodity prices and the ultimate financial well-being of U.S. farmers. And as the government supports farm income during low price periods, a competitive U.S. agriculture has a major impact on the federal budget.

Evenson concentrates on technological competitiveness, showing the diminished comparative advantage of U.S. agriculture relative to other suppliers to the world market. The United States is still on the vanguard of new technology development, yet the data show that other nations are growing of their influence as sources of inventions.

Hunger remains a global problem yet the overall supply of food remains sufficient for the population. The problem is that resources are badly distributed both among and within nation states. Efforts to help Third World farmers to be more effective can be augmented by biotechnological advances. Brady expresses the cautious excitement that underlies the future for world food production. A heightened responsiveness in agricultural research has been fueled by the rapid diffusion of enhanced technical capability to agricultural scientists around the globe.

Geisler and DuPuis examine the role of institutional actors in mediating the impacts of new technology as outlined by classical diffusion theory. They call for a new paradigm that places greater emphasis on farmer participation and equity, as well as on adaptation rather than adoption.

The comprehensive shifts that will be based on the potential of biotechnology can only be hinted at by the chapters in this volume. What is clear, however, is that the biotechnology revolution is upon us and that technical change must be informed by a broader understanding of the social and institutional consequences of the choices to be made.

REFERENCES

Alexander, M. 1985. Ecological consequences reducing the uncertainties, Issues in Science and Technology 1:58-68.

Arntzen, C. J. 1984. Cutting edge technologies. Washington, D.C.: National Academy of Engineering, pp. 52-61.

Brill, W. J. 1985. Regulation of biotechnology, Science 227:381-84.

Brown, L. A. 1981. Innovation diffusion. London and New York: Methuen.

Buttel, F. H. 1985. The land-grant system: a sociological perspective on value conflicts and ethical issues. Agriculture and Human Values 6:78-95.

Buttel, F. H. 1986. Biotechnology and agricultural research policy: emergent issues. In K. A. Dahlberg (ed.), New directions for agriculture and agricultural research. Totowa, NJ: Rowman and Allenheld.

Buttel, F. H., Cowan, J. T., Kenney, M., and Kloppenburg, J., Jr. 1984. The political economy of agribusiness reorganization and industry-university relationships. Research in Rural Sociology and Development 1:315-43.

Carson, R. 1962. Silent spring. Boston: Houghton-Mifflin.

Cochrane, W. W. 1979. The development of American agriculture. University of Minnesota Press: Minnesota.

Crott, R. 1986. The impact of isoglucose on the international sugar market. In S. Jacobsson, A. Jamison and H. Rothman (eds.), The biological challenge. Cambridge: Great Britain at the University Press, pp. 96-123.

Doyle, Jack. 1985. Altered harvest. New York: Viking Penguin.

Evenson, R. E. 1984. Scientific Credibility and Applied Agricultural Research. Paper presented at the AAAS Annual Meeting, New York.

Fallert R., McCuckin, T., Betts, C., and Brunner, G. 1987. BST and the dairy industry. Washington, DC: USDA-ERS-AER report 570.

Hanway, D. C. 1978. Crops and Soils Magazine 30:5-6.

Hess, C. E. et al. 1987. Agricultural biotechnology. Washington, DC: National Academy Press.

Jaffe, G. A. 1987. Inadequacies in the federal regulation of biotechnology. Harvard Environmental Law Review 11:491-550.

Jeffries, P. 1987. Use of mycorrhizae in agriculture. Critical Reviews in Biotechnology 5:319-57.

Kalter, R. J. et al. 1984. Biotechnology and the dairy industry: production costs and commercial potential of bovine growth hormone. Department of Agricultural Economics A-E Research 84-22. Ithaca, New York: Cornell University.

Kenney, M. 1986. Biotechnology: the university-industrial complex. New Haven and London: Yale University Press.

Khachatourians, R. 1986. Production and use of biological pest control agents. Trends in Biotechnology 5:120-124.

Kloppenburg, J., and Kleinman, D. L. 1987. Seeds of struggle: the geopolitics of genetic resources.

Technology Review 90:46-53.

Kloppenburg, J. 1988. First the seed; the political economy of plant biotechnology. New York: Cambridge University Press.

Kvistigaard, M., and Olsen, A. M. 1986. Biotechnology and environment: a technology assessment of environmental applications and impacts of biotechnology. Trends in Biotechnology 4:112-15.

Lienhard, J. H. 1985. Some ideas about growth and quality in technology. Technological Forecasting and Social Change 27:265-281.

Miller, J. D. 1983. Scientific literacy: a conceptual and empirical review. Daedalus 112:29-48.

Molnar, J. J., and Kinnucan, H. 1985. Biotechnology and the small farm: implications of an emerging trend. In T. T. Williams (ed.) Strategy for survival of small farmers, Tuskegee, AL: Tuskegee University Human Resources Development Center.

Molnar, J.J., Kinnucan, H., and Hatch, U. 1986. Anticipating the impacts of biotechnology on agriculture: a review and synthesis. In H. M. LeBaron et al. (eds.), Biotechnology in agricultural chemistry. Washington, DC: American Chemistry Society, pp. 254-67.

National Academy of Sciences. 1982. Genetic engineering of plants. Washington, D.C.: National Academy of Science.

Office of Technology Assessment. 1982a. Genetic technology, a new frontier. Boulder, CO: Westview Press.

Office of Technology Assessment. 1982b. Impacts of technology on U.S. cropland and rangeland productivity. Washington, D.C.: U.S. Congress, Office of Technology Assessment.

Office of Technology Assessment. 1984. Commercial biotechnology: an international analysis. Washington, D.C.: U.S. Congress, Office of Technology Assessment.

Perrow, C. 1984. Normal accidents: living with high-risk technologies. New York: Basic Books, Inc.

Prentis, Steve. 1984. Biotechnology: a new industrial revolution. New York: George Brazilier.

Rees, J. 1985. Natural resources: allocation, economics and policy. London and New York: Methuen.

Rogers, E. M. 1982. Diffusion of innovations. New York: The Free Press:

18

Shemilt, L. W. (ed.) 1982. <u>Chemistry and world food supplies: the new frontiers, chemrawn II</u>. Canada: Pergamon Press.

Thurow, L. C. 1985. <u>Resources</u> 80:5-9.

Tonalski, W., and Rothman, H. 1986. Enzyme technology. In S. Jacobsson, A. Jamison, and H. Rothman (eds.), <u>The biotechnological challenge</u>. Cambridge: Great Britain at the University Press, pp. 37-76.

van den Doel, K., and Junne, G. 1986. Product substitution through biotechnology: impact on the Third World. <u>Trends in Biotechnology</u> 4:88-90.

Williams, S. B. 1986. Utility product alter protection for plant varieties. <u>Trends in Biotechnology</u> 5:33-39.

Yoxen, E. 1986. The social impact of biotechnology. <u>Trends in Biotechnology</u> 4:86-88.

_____. 1984. <u>The gene business.</u> New York: Harper and Row.

Part I

Institutional Issues

2. The Changing Division of Labor Between the University and Industry: The Case of Agricultural Biotechnology

Recently the president of a large land grant university noted, "Our society has moved into the era of high technology. In the next two decades before the turn of the century, we will see more technological changes than we have experienced over the entire history of our nation. It promises to be one of the most exciting and challenging times in the history of mankind." He further observed, "Biotechnology, genetic manipulation, and engineering research will have tremendous impact on the crops and animals we grow for food, affecting agriculture in ways never before dreamed possible" (personal interview 1985). Although this optimistic view is not shared by everyone, the promise of the new biotechnologies for agricultural research has attracted new interest and substantial investment in agricultural research from the agribusiness community, chemical industry, venture capitalists, state and federal governments, and the land grant universities. A rich variety of interdisciplinary centers, institutes, laboratories, research parks, and corporations have been created at American universities to pursue agricultural biotechnology. Since World War II, universities generally have adapted successfully to these entities operating outside the traditional departmental structure. (The 1984-1985 Research Center Directory lists over 7500 non-profit research organizations, a majority of which are university related.) However, the emergence of recent affiliated institutions is causing some uneasiness.

New research centers focus on the major growth areas of biotechnology and information technology and tend to be larger and better financed than their precursors. In addition, these new research institutions are often being established and operated by mixed partnerships. This has renewed concern about the appropriate roles for public and private research (Buttel et al. 1986; Doyle 1985; Hansen et al. 1986; Kenney 1986; Kloppenburg 1987). Guidelines for collaboration have been suggested for publication rights, patent ownership, licensing, copyrights, confidentiality agreements, research units, faculty consultants, faculty entrepreneurs, international agreements, and the sharing of personnel and equipment. In the area of agricultural biotechnology, these issues prompted the land grant Experiment Station Committee on Policy (ESCOP) to express its concerns publicly and to develop guidelines to deal with these questions. The National Association of State Universities and Land Grant Colleges' Division of Agriculture also established a blue ribbon committee to explore a range of issues and policies, including the legal framework for scientific inquiry at public universities and guidelines for the development of university-industry research contracts (National Association of State Universities and Land-Grant Colleges 1983, 1984, 1985).

While partnerships between university and industry have existed for several decades, these new types of university-industry relationships in biotechnology are more varied, more aggressive, and more experimental. They include: large grants and contracts between companies and universities in exchange for patent rights and exclusive licenses to discoveries; programs and centers organized at major universities with industrial funds that give participating private firms privileged access to university resources and a role in shaping research agendas; professors, particularly in the biomedical sciences, serving in extensive consulting capacities on scientific advisory boards or in managerial positions of biotechnology firms; faculty receiving research funding from private corporations in which they hold significant equity; faculty setting up their own biotechnology firms; and public universities establishing for profit corporations (such as Neogen at Michigan State) to develop and market innovations arising from research. Finally, state governments in many parts of the country are establishing R&D centers linking industry to universities with a view toward promoting economic development (Krimsky 1984; Kenney 1986; Glazer 1986; Jaschik 1986). Twenty-seven states have established centers or programs devoted specifically to research on areas directly related to biotechnology in the last five years. New York allocated $34 million in 1986 with 17 other states reporting allocations in excess of one million. These state-supported biotechnology programs rely primarily on their universities for the design and performance of the research and training. In many states, university-industry collaboration is a condition for qualifying for research funds (Freeman 1987).

These experimental relationships are raising new issues for universities. Does corporate penetration into academic science distort traditional values of basic research and alter the atmosphere, working climate, and free communication of academic departments? There is also concern by universities to assure that research projects are genuinely originated by faculty members and not adopted as a result of outside pressure, either implicit or explicit. Related to this is the issue of the nature of the research agendas. It has been proposed that if a sufficiently large and influential number of academic scientists and engineers became involved with industry, a whole range of research agendas, traditionally part of the university community, might be abandoned. Furthermore, the scientific community could become desensitized to the social impacts of the biotechnology research. Certain research focused on problems related to the commercial-industrial applications of the particular area of science-engineering could be neglected entirely (Krimsky 1984). A further issue involves the potential conflict of interest between industry and university agendas. Moreover, the new research centers generate the additional conflict of divided commitments and loyalties created by joint appointments at both the university and the center. Universities are concerned that they maintain control of appointments to these centers. Beyond the boundaries of a single university is the issue of these new centers conferring a high tech advantage on a particular firm or a single university.

In addition to the important potential impacts of biotechnology on the organization and conduct of agricultural research, strong concerns have arisen from critics who question the possible ecological consequences of this research. The economic and environmental benefits expected to occur from agricultural use of recombinant organisms are great. They include (1) the creation of plants resistant to a wide array of pests leading to possible decrease in the use of chemical fungicides and insecticides, (2) the development of plants that utilize sunlight and fertilizers more efficiently, and (3) introduction of new genes and the consequent increase in genetic variability for the future (Netzer 1987a; 1987b; Fischhoff et al. 1987). However, they may include the creation of plants resistant to particular pesticides and herbicides and which are dependent for maximum productivity on specific packages of inputs. This may increase farmer input costs and further exacerbate environmental impacts of increased chemical use in food production. In addition, there have been particular safety concerns regarding the introduction of new organisms and the unexpected consequences that might occur such as the production of a toxic secondary metabolite, protein toxin, or undesired self-perpetuation and spread of the organism (Brill 1985). Some critics of the research have successfully sought court injunctions to prevent research that entailed the release of new microbes in a natural setting.

With these recent developments as a context, this chapter examines the impact of agricultural biotechnology on the organization and conduct of public and private sector agricultural research.[1] As noted by sociologists of science, R&D organizations not only provide the facilities for their members but also the environment that may either stimulate or inhibit the scientists' communications, performance, and research, which ultimately shapes the products and their impact in the broader society (Pelz and Andrews 1976; Andrews 1979; Busch and Lacy 1983; Lacy and Busch 1983).

BACKGROUND

In the past ten years dramatic new developments in the ability to select and manipulate genetic material have generated a new basic science frontier in the public research sector and ignited unprecedented interest in the industrial use of living organisms. Howard Schneiderman, senior vice president for life science research at Monsanto Corporation, indicates that this new biotechnology is "absolutely a global market" and one that some experts believe will reach \$100 billion in sales by the twenty-first century (Schrage and Henderson 1984).

After the success of the first directed insertion of genetic material in a host microorganism in 1973, researchers have recognized the opportunities for knowledge in basic molecular biology and the commercial potential in pharmaceuticals, animal and plant agriculture, specialty chemicals and food additives, environmental areas, commodity chemical, energy production, and bioelectronics. The first expression of a gene cloned from a different species in bacteria followed in 1974. In 1976 Genentech, the first firm to exploit rDNA technology was founded. Other major events in the last decade have included the establishment of over 80 new biotechnology firms by the end of 1981; the 1982 approval of the first rDNA pharmaceutical product (human insulin) for use in the United States and the United Kingdom; and the first expression of a plant gene in a plant of a different species in 1983.

Simultaneously, changes have been occurring in the patent laws and court interpretations that have facilitated if not stimulated the development of agricultural biotechnology. In 1970, Congress passed the Plant Variety Protection Act, which provided patent-like protection to new, distinct, uniform, and stable varieties of plants that were reproduced sexually. More important, however, for the new biotechnologies was the landmark U.S. Supreme Court case _Diamond_ vs. _Chakrabarty_ (1980), which provided complete patent protection for genetically engineered microorganisms. This has been followed by the 1985 Patent Office ruling that utility patents could be granted for novel plants (_Ex parte Hibberd_ 1985) and by the patent application decision _In re Allen_ that genetically altered oysters were patentable subject matter. In contrast to PVPA, these recent decisions provide exclusive patent protection for higher life forms as long as the organism results from human ingenuity and intervention (Merges 1987; Buttel and Belsky 1987).

The biological developments and legal decisions have been paralleled by major financial investments in these new techniques both in the United States and in countries such as Japan, the Federal Republic of Germany, the United Kingdom, Switzerland, and France. In just a decade biotechnology has moved from a test tube experiment to a high industrial priority of governments. Japan's Ministry of International Trade and Industry, which has guided development efforts in automobiles and computers, has designated biotechnology a strategic industry. In 1983, $100 million was allocated to Japanese firms for biotechnology research and development (OTA 1984). In West Germany, three federally sponsored biotechnology research institutes joined with industry in investing an estimated $280 million in this research in 1985.

In the United States, the federal government expenditure for basic research in biotechnology -- over $600 million annually in recent years -- is the largest in the world. (This has been more than matched by the private sector investments to commercialize biotechnology, which first exceeded one billion dollars in 1983.) In addition, states entered the funding picture in the early 1980s. As recently as 1980, only 10 states had programs promoting high technology with minimal funds devoted to this purpose. By 1986, forty states were funding high technology with thirty-three states investing in the development of biotechnology. State allocations in biotechnology in 1986 and 1987 amounted to nearly $100 million each year. Moreover, both states and the federal government are investing heavily in education and training in biotechnology. The National Institutes of Health, for example, is spending about $60 to $70 million annually to train research scientists in the skills of biotechnology (Makulowich 1987).

In 1984, the amount of federally funded research in agricultural biotechnology was approximately $70 million, and the Department of Agriculture expended approximately $50 million for biotechnology research (GAO 1985). In addition, private industry has been investing heavily in public sector agricultural research. In the 1983 academic year, agribusiness contributed an estimated $40 million to university research for bioengineering research, and in 1984 biotechnology companies spent about $120 million in grants and contracts to universities. This funding represented approximately 16-24 percent of the public sector universities' total funds expended for biotechnology in 1983 and 1984 compared to an average of 3-5 percent that industry provides for all research funds expended in institutions of higher education (Blumenthal et al. 1986a; 1986b). Therefore, industry provides a much larger proportion of the funds available for university research in biotechnology than in other fields. Indeed, in Texas and Indiana, the percent of private sector research support for biotechnology exceeded 20 percent during the mid 1980s (NASULGC 1983, 1985). This corporate support for university agricultural biotechnology is expected to increase as much as tenfold by 1990 (Matthiessen and Kohn 1984).

The private sector has also stepped up its own in-house research. In the United States, two distinct groups are pursuing commercial applications. The first group is represented by the established companies that are generally process-oriented and multiproduct companies in traditional industrial sectors, such as pharmaceuticals, energy, chemicals, and food processing. This includes such large established firms as E.J. du Pont de Nemours and Co. Inc., W.R. Grace and Co., Monsanto, Eli Lilly and Co., Merck, Allied Chemical Corp., and ARCO.

The second group is represented by the new biotechnology firms that are entrepreneurial ventures started to commercialize innovations in biotechnology and founded for the most part since 1976. They are almost exclusively an American phenomenon. Substantial venture capital has been generated to fund the startup of more than 100 new biotechnology firms (NBFs). Between March and July of 1983, 23 NBFs raised about $450 million. In addition, corporate equity investment in NBFs, although now diminishing, has totaled over $350 million from 1977 to 1983.

According to a 1984 Office of Technology Assessment report, 62 percent (135) of the 219 U.S. companies pursuing applications of biotechnology are working in the area of pharmaceuticals, while 28 percent are involved in animal agriculture and 24 percent are involved in plant agriculture. By the end of 1984 the number of biotechnology companies in the U.S. had grown to 270. The Genetic Engineering News "1984 Guide to Biotechnology Companies" contained 322 companies from 16 countries, an increase of more than 100 companies in a year. While 193 reported work in diagnostics and 144 reported work in pharmaceuticals, there were also 114 companies pursuing work in agriculture. A 1985 survey of the agriculture biotechnology firms indicated they employed more than a thousand molecular biologists and invested over $200 million in agricultural biotechnology research and development in 1984. Most recently,

the Genetic Engineering News "Guide to Biotechnology
Companies" contained 522 companies from 24 countries
with 379 from the United States. Despite some consolidation and decline in venture capital, biotechnology
continues to be a major growth industry.

As noted above, the largest number of companies
and the first applications of biotechnology have
emerged in the area of pharmaceuticals. Several factors explain this development: (1) rDNA and monoclonal
antibodies technologies were developed with public
funds directed toward biomedical research; (2) the
pharmaceutical companies have had years of experience
with biological production methods and this experience
has enabled them to take advantage of the new technologies; (3) there are no competing production methods to
biological methods for some pharmaceutical products
such as large polypeptides and antibiotics that might
inhibit the application of biotechnology to their production; (4) pharmaceuticals are profitable products
because they are low volume and high-value-added products. In the United States about 78 NBFs and 57
established companies are reported to be using
biotechnology to pursue pharmaceutical product and
process development.

In the area of animal agriculture, recent developments such as the advent of biotechnology, rising
industrialization of animal agriculture, and changing
dietary habits in foreign countries have increased
demands for improvements in old products as well as for
completely new products. The major segments of the
industry where biotechnology is seen as a potential
contributor are in diagnostic products, growth promo-
tants, and vaccines. Of the 61 companies known to be
pursing animal related applications of biotechnology,
34 (56 percent) are NBFs. Interestingly, most of the
established U.S. companies have made relatively small
investments in this area, equal to or less than the
investments in animal health by most of the leading
NBFs. In contrast to human pharmaceutical products,
animal vaccines and diagnostic products are in many
cases being developed independently by NBFs without the
support of established U.S. companies (OTA 1984).

The plant agricultural area includes researchers
and firms that are engaged in work to (1) modify specific plant characteristics (e.g., tolerance to stress,
nutritional content, yield, resistance to pesticides
and herbicides, or photosynthesis efficiency), and (2)

to modify traits of microorganisms that could be important to plant agriculture (e.g., nitrogen fixation, disease suppression, or frost resistance) (Netzer 1987a, 1987b; Fishhoff et al. 1987). Although there are 22 NBFs among the 52 diverse firms in this area, applications of biotechnology to plants appears to be dominated by the established companies in oil, chemicals, food, and pharmaceuticals (e.g., American Cyanamid, Dow, Allied, DuPont, Monsanto, ARCO, Shell, Ciba-Geigy, Rohm & Haas, Sohio). One route by which some established U.S. companies have entered the plant agricultural field is through the acquisition of seed companies (Kenney 1986; Buttel et al. 1986; Kloppenburg, 1987).

Despite the large influx in these new venture capital biotechnology firms over the last decade, the momentum of the biotechnology industry has shifted to the large established chemical and pharmaceutical firms. These firms are increasing their already significant investments in research. They are also increasing their efforts to scale up for worldwide production and forming close ties with the smaller NBFs to gain access to their technology. At the same time current industry conditions are encouraging consolidation and mergers.

First, outside sources of funding such as the public equity market, research and development limited partnerships, venture capital, and corporate funding are decreasing dramatically. From 1981 to 1983 investors provided nearly $2 billion to the industry. In contrast, support in 1984 was only a small fraction of that level (especially for smaller, less well capitalized firms) and even less in 1985. Concurrently, approximately three-quarters of all public biotechnology companies continue to operate at a loss and most of the timetables for delivery of products have not been optimistic. Consequently, promised commercial applications have not been realized as anticipated and there has been a lack of internally generated cash flow from product sales. Furthermore, many of the products under development appear to duplicate research efforts at other companies in the industry. In addition, in firms such as Biogen, Cetus, Collaborative Research, and Genex there has been a transition in management from the founding scientists to outside managers (Miller 1985). Some suggest that the removal of management from its founders increases the likelihood of a sale of

the company. Finally, although many companies have
some of the capabilities required to market a new
product (i.e., research, development, formulations,
scale-up, preclinical and clinical testing, manufacturing, and marketing), most do not have all of these
prerequisites in place.

Over the next few years it is very likely that a
large number of these biotechnology companies will
cease to do business as independent enterprises. Nevertheless, the activity and impact of the private sector in biotechnology research is likely to continue to
increase. By the end of the century, sales just from
bioengineering farm products could amount to $100 bil-
lion, according to the Policy Research Corporation in
Chicago. "We are entering a new era," says Orville
Bentley, the USDA Assistant Secretary for Science and
Education. "The potential is practically unlimited"
(Matthiessen and Kohn 1984).

THE INSTITUTIONAL CONTEXT

A major determinant of the extent to which the new
biotechnologies will have a significant impact on food,
agriculture, and agricultural research is the institutional structure and climate wherein biotechnology is
undertaken. A major feature of this institutional
context is the division of labor between the public
sector institutions and private industry. One point
repeatedly made by administrators in both the public
and private sectors is that the increase in biotechnological research and development is likely to have
significant impacts on this division of labor, which
will have important ramifications for the future of
both public and private R&D concerning food.

Traditionally, public sector research institutions
have been "responsible" for longer-term, higher-risk
R&D. Private sector institutions have focused upon
shorter-term, profit-oriented projects. Further, the
mission-oriented nature of public research institutions
has tended to emphasize research in service to a number
of publics or client groups: farmers, processors,
inputs suppliers, and consumers. Alternatively, the
R&D enterprise in the private sector has been geared
specifically toward the interests of the firm. A consequence of this is that the criteria for the choice of

research problems and the development of technologies, management strategies, etc., have differed between the public and private sector. Private sector R&D, which is geared toward profitable products, has tended to focus upon efforts to (1) truncate the time and space within which production or processing take place and (2) increase the precision of techniques or technologies for production or processing. These factors affect profitability of production and processing, and hence, are not unexpected primary considerations by private firms. What is interesting is that these criteria motivated only some of the R&D performed in the public sector, namely, research aimed specifically at serving the production or processing segments of "the public."

In contrast, there are many other goals highly emphasized in public sector R&D: quality of rural communities, long term sustainability of our foods, conservation, and nutrition to mention only a few. The most important criterion in public R&D has been simply to serve the public. The division of labor has been, until the present, fairly clear (Lacy and Busch 1984). Recently, however, the public sector has begun to turn its attention toward "private sector" goals in attempts to truncate time and space in production, as well as to increase product premiums.

Changes in both the division of labor between the private and public sector, as well as in the criteria utilized in selecting research and development projects have been brought about largely because of the broader economics of agricultural production and processing. The trend in agriculture as a whole has been toward increased industrialization and concentration. The situation is analogous to other industries such as that of handicrafts in the late eighteenth century. The factory system truncated time and space and gave greater precision to the production process. Consequently, household handicraft production was gradually eclipsed by capitalist factory production (Sohn-Rethel 1978).

It was in the food processing sector that industrialization and concentration first took place in the food system. As early as 1785, Oliver Evans had designed a fully-automated, continuous process flour mill. By 1789 bread-kneeding machines were in operation in Genoa. Hog-slaughtering was a partially automated process as early as the 1830s. Later in the nineteenth century, pasteurization brought regularity and mass production to milk and cheese production. More recently, poultry production has been fully industrialized by a massive increase in flock size and assembly line processing techniques (Burkhardt et al. 1986; Cochrane 1979; Giedion 1975).

Close behind the development of processing followed the rise of the farm equipment industry. As early as the 1830s, stationary steam engines were in use on American farms. By 1880, four-fifths of all wheat was reaped by machine. The advent of the tractor at the turn of the century further increased the area that could be cultivated by a single man (Giedion 1975).

Clearly, then, some segments of the agricultural production and processing system industrialized earlier. Farm machinery became concentrated relatively early, while concentration in vegetable processing and farming is still underway. The interesting point is that R&D designed to increase industrialization and concentration through greater spatial-temporal truncation and increased precision only generates a marginal return on the investment in the R&D if industrialization and concentration are in their incipient stages. Thus, greater attention is being paid to "new" technological developments in plant and animal improvement (contributing toward industrialization of farm production) than to biotechnology in food processing (except in special cases), at least by private industry.

Research and development in the private sector has been and is still motivated by the fact that returns to research may be captured by those doing (or paying for) the research. Patent laws further insure that new processing techniques and machinery yield a satisfactory rate of return. In contrast, returns to research in the public sector cannot be captured in a similar fashion, and indeed, there has been no motivation for doing so. Besides its "mission," public agricultural research had the luxury of "unspoiled" or at least not necessarily commercially-oriented R&D.

The funding structure for agricultural research facilitated the high-risk and/or long-term developmental projects of the land grant system. Hatch funds or federal "formula funding" and related state funds accounted for the major source of an institution's support and USDA or state grants and contracts have been the major sources of supplemental support for individual scientists (Busch and Lacy 1983; Zuiches 1986). This situation is changing. Hatch or formula funding is now a relatively small portion of total funds. USDA and other competitive grants, grants from commodity associations, state grants and contracts, and more recently large corporate grants are replacing the traditional "institutional" funding for the experiment stations.

The shift from institutional funding is partly the result of the eclipsing of public agricultural research at the federal level by basic science, medical, and military research. While agricultural research was the most visible of all federal research in 1940, today it is just a small portion of the total research budget. In addition, there is increased pressure on public institutions for accountability in research, teaching, and extension. Where colleges of agriculture report directly to and are directly funded by their respective state legislatures, program justifications are a funda-mental component of the annual budgetary request. In many states, population shifts toward urban areas have redistributed the rural/urban balance in the legisla-tures, which makes traditional programs more difficult to defend. Moreover, colleges of agriculture have become submerged in large state universities.

The population shift toward urban areas has also contributed to a major shift in the purpose and func-tion of the Land Grant Universities (LGU). The LGU system was initially set up in 1862 to counter the growing elitism of private universities. The LGU had a mission orientation to serve the needs of rural people. Its purpose was to educate the rural masses and to improve the quality of rural life by doing research on the pressing social and agricultural problems of the day. Scientists were rewarded for being good teachers and for doing applied research that addressed social problems. Over time, however, the LGU seems to have lost this component of its mission orientation as the rural population steadily declined in numbers and power.

This decline, coupled with the pervasive views that rural areas are more "backward" than urban areas and that "basic" science is more interesting and "bet-ter" than applied science, encouraged the LGU system to shift its focus more toward high technology and basic science. Furthermore, as private industry has increas-ingly become involved in agriculture in general, and crop breeding in particular, LGUs have felt pressure to focus on basic science and to leave applied work -- such as variety development -- to private industry. Private industry argues that the public sector should not be competing with the private sector.

Finally, criticism of the agricultural research establishment (in the Pound and Winrock reports) as behind the times has also spurred the shift toward more glamorous basic science. Indeed, colleges of agriculture increasingly emphasize the academic criteria of basic science such as number of publications in peer-reviewed journals (Busch and Lacy 1983). When evaluating agricultural researchers for promotion, teaching and applied work (such as variety development) has been de-emphasized. Rethinking of the traditional goals of the experiment station, and of public institutions in general, seems to have occurred (Busch and Lacy 1986).

Changes have also occurred in the private sector. Before the second World War, private firms involved in agriculture were confined largely to the manufacture of farm machinery and the processing of farm products. With the development of the agrichemical industry, in part as a result of the war, private sector interest in chemical R&D increased considerably. More recently, passage of the Plant Variety Protection Act (PVPA) in 1970 allowed the patenting of most sexually reproducing plant varieties, which made variety development more profitable and led to major changes in the seed industry. Large multinationals, primarily agrichemical and pharmaceutical companies, rapidly began buying seed companies (Kloppenburg 1987). By 1980, five companies, all with less than ten years experience in plant breeding, owned 30 percent of all U.S. plant variety patents. In fact, 72 percent of all the plant patents issued went to six crops where a handful of multinational firms clearly dominate the industry (Mooney 1980; 1983). The net effect of the PVPA was to stimulate capital investment and economic concentration in the seed industry.

A similar effect has occurred as a consequence of the 1980 Supreme Court decision permitting the patenting of novel life forms (Diamond vs. Chakrabarty 1980). After this decision, patent activity from both the public and private sectors in genetic engineering increased dramatically. Indeed, since 1980 patent activity in genetic engineering has increased at over twice the rate of other technologies compared with 1974-1980 rates (NAS 1984).

There have been other changes in the private sector as well. Partly as the result of increased costs of purchased inputs, and partly as a result of surpluses for many agricultural commodities, the number of intermediate-sized farms has rapidly declined in the past two decades. This has led the client groups for public agricultural research to shift from smaller family farms to large farming enterprises. Furthermore, the market shares of large processing companies such as Campbells and Hunts, and millers such as ITT, have increased significantly. In general, the nature of the "product" coming out of public agricultural research institutions has had to change somewhat -- albeit slowly in some cases -- toward the new "publics" served by these institutions.

The advent of the new biotechnologies has stimulated the interest of seed companies, multinational chemical firms, large farm enterprises, and food processors. In part, this is because the new biotechnologies have the potential to further truncate time and space, as well as to increase the precision of plant and animal breeding programs and the efficiency of processing. Until recently, the very biological character of agricultural products acted as a barrier to the rationalization of their improvement. The new biotechnologies, however, hold the potential for allowing greater control over production and processing. For example, bovine growth hormone (BST) appears capable of increasing per cow milk production by 10-40 percent. Only one or two plants would be able to supply all the BST needed by the nation's dairy farmers. Widespread adoption of BST is likely to lead to increased herd size (Kalter et al. 1984) and perhaps a geographical redistribution of the dairy industry as well. Similarly, the development of flavor enhancers, new fermentation techniques, and even the in vitro production of certain types of processed foods is now within the realm of possibility. There is even talk of the replacement of agriculture by genetically-engineered synthetic foodstuffs (Rogoff and Rawlins 1986).

The private sector, from all appearances, is eager to renegotiate the division of labor with the public sector, or at least to have greater input into research agendas. Profitable products may result from university-industry negotiations. Indeed, private industry has increasingly looked to the university, both as a source of trained scientists and of technologies that might be useful in the pursuit of profits. With the increased difficulty of obtaining public funding, experiment stations have accepted industry's largess. Private contracts and grants now account for a greater share of the funding base for particular projects and, indeed, for entire programs (Blumenthal et al. 1986a; 1986b). As noted earlier, 46 percent of all firms in the biotechnology industry support biotechnology research in universities ($120 million in 1984). Furthermore, industrially sponsored university research in biotechnology is approximately 16-24 percent as compared to the average of 4-5 percent of overall campus research supported by industry. In addition to the trend of larger industry grants, there tend to be more "strings" attached, including delaying publication of research results or demanding exclusive licensing of any patentable products or processes. Experiment stations have begun to tailor certain Ph.D. programs to industry needs and have accepted the challenge to produce goods and services for industry.

Accountability, demands for productivity, and funding shortages (or at least great competition for funding) appear to have encouraged station directors to view favorably the new biotechnologies. Given the economic and structural aspects as well as the inherent importance of scientific knowledge in this area, it is likely that public universities will seek more private sector grants and hire more young faculty to do predominantly biotechnological research. Private industry certainly seems to be eager to provide capital for biotechnology projects. Thus, biotechnology is a prime candidate for receiving a high priority among the various goals that an experiment station can pursue. Moreover, various science-policy groups at the federal government level (NRC 1987) have targeted biotechnology as the first priority in goals for scientific research. These factors only reinforce that biotechnology is the key to the future of agricultural research and development.

THE IMPACTS

The social and scientific impacts of the new biotechnologies are occurring, and will continue to occur, within this complex institutional milieu. Indeed, part of our assessment of these technologies is that they are in large measure being used by private sector corporations, experiment stations, USDA-ARS, and the universities as institutional tools in order to secure public funding, marketable products, and ultimately, the existence of the public research institution and private funding for that continued existence. However, it is unlikely that university-industry relationships in biotechnology will significantly meet the unmet operating capital needs of the universities. Furthermore, there are more specific impacts with respect to particular functions of experiment stations and the interests of scientists, administrators, farmers, processors, and ultimately, the public.

The impacts of biotechnologies are clearest in the case of animal and plant improvement programs, although there are analogous changes occurring in food science as well. For example, in the area of plant agriculture, the potential of the new biotechnologies appears to have accelerated a shift that began in the 1920s of varietal breeding from the public to the private sector. Until 1923, USDA regularly distributed free seed, collected from around the world, to farmers who wanted to test the seeds on U.S. soils. As the techniques and effectiveness of plant breeding improved, USDA discontinued the distribution of free seeds. Instead, state agricultural experiment stations (SAES) began to disseminate new varieties developed by their breeders to seed companies who multiplied and sold the new varieties to farmers. The development of hybrid corn shifted the division of labor again. Experiment stations gradually shifted to the development of parent material and significantly reduced the production of finished products for hybrid crops. With the growth of the seed development (as opposed to multiplication) industry, came increased pressure on the public sector to cease producing finished varieties. In 1923 the American Seed Trade Association (ASTA) began to encourage this development with most of the pressure coming from the larger companies. A recent survey of ASTA members revealed that small and medium-sized companies placed more importance on hybrid and variety development by USDA/ARS and SAES than do large companies. Comparable results were found for companies with no R&D vs. companies with R&D (ASTA 1984).

Large seed companies generally believe that experiment stations should do basic research and produce enhanced germplasm that can be developed into new varieties by the seed companies. The new biotechnologies are viewed as the tools for doing this basic research. Both views are frequently echoed by experiment station directors. Pressure on public sector scientists from industry to abandon varietal breeding, pressure from administrators for greater productivity, as well as the lure of large amounts of private money for biotechnology research has led to a change in disciplines in the SAES. A recent survey indicated that an additional 151 full-time equivalent scientists (FTE) were to be added in the biotechnology disciplines (plant and animal sciences as well as food science) between 1985 and 1987 (NASULGC 1985). This followed the previous addition of 90 molecular biologists between 1982-1984 bringing the total to 373 FTEs representing 6 percent of the total SAES faculty FTEs. Interviews with SAES directors indicated that many of the molecular biology positions were obtained by reducing the scope of conventional breeding programs. Retiring plant breeders were replaced by molecular biologists. A similar process appears to have occurred within the Agricultural Research Service (ARS). The six year plan for 1984-1990 called for ARS scientists to emphasize "basic" science (i.e. molecular biology) and germplasm enhancement and to de-emphasize varietal breeding and some small commodity programs (USDA-ARS 1983). In fact, the plan called for a reduction of two million dollars for breeding of horticultural crops.

A survey of public breeding programs (in the SAES) of horticultural and sugar crops reveals a projected 22-55 percent reduction in the number of programs and a 13-39 percent reduction in FTEs by 1990 (Hansen et al. 1986). A similar survey of public breeding programs for field crops reveals a projected 9-29 percent reduction in full-time positions by 1990. Already between 1982-1984 there was a decline of 17 percent among plant breeders and 21 percent among animal breeders for a loss of eighty plant breeding positions and thirty-five animal breeding positions (NASULGC 1983; 1985). Much of this reduction will be due to the termination of smaller programs, especially in the minor crops.

These developments raise a number of important questions. First, who will do the breeding for the minor crops? The problem of minor crops and minor

animals (i.e., sheep or goats) is very similar to that
of orphan drugs. The market is either limited or
insufficiently profitable. As a result, the private
sector invests most heavily in major crops and also
pressures the public sector to focus on these crops
through grants, contracts, consulting, and even legislation. The public sector also focuses on major crops
as a result of state and national funding practices.
Furthermore, with public funds available for biotechnology programs involving only a few crops, research is
likely to follow the interests of the most powerful
commodity associations that tend to represent the major
crops, rather than being used to increase the role of
minor crops or the number of food crops available for
human use. However, some private sector breeders are
aware of the problem and want the public sector to
continue minor crop breeding (Kalton 1984; Hansen et
al. 1986).

Another equally important question concerns the
training of the next generations of plant breeders.
There is a growing demand for breeders in the private
sector (Kalton 1984), both in seed companies and in
biotech firms or departments in large corporations.
Even varieties and hybrids developed in tissue culture
laboratories need to be grown out in the field. Breeding programs are deemphasized as universities emphasize
biotechnology or basic research. This may severely
limit the personnel available to train the future
breeders. Plant breeders require practical experience
to complement learning from books. This experience is
obtained in the process of breeding and supervised by
an experienced breeder.

Similar points can be made about animal sciences,
particularly husbandry. Increased emphasis on bioengi-
neered hormones, foodstuffs, and even altered distributions of fat, protein, and water content in animals may
have significant impacts on how the animal sciences are
practiced, and ultimately, on the products available
for processing. Although the impacts on
university-industry collaborations may not be as
pronounced in the animal sciences and meat processing
industry -- given the already highly industrialized and
concentrated nature of this industry -- there will
nonetheless be some impacts. As we noted, a major
impact will probably be the further tailoring of
public-sector research and training to very specific
industry demands.

Additional impacts can be projected for the structure of public institutional agricultural research. In
particular, a significant increase is likely in the
concentration of scientific talent at a small number of
public and private institutions. For example, every
state has been able to afford a conventional plant
breeding program. But not every state can afford to
have a comprehensive plant biotechnology program. In-
strumentation costs are particularly expensive
($125,000-$150,000 just for initial lab set up)
(NASULGC 1983). In a 1985 survey of twenty-nine companies doing agricultural biotechnology, average annual
total support per principal investigator was 161,000
with four companies averaging over a quarter million
dollars per scientist (NASULGC 1985). Minimum funding
levels of $100,000/year/grant are necessary, according
to SAES directors, for adequate funding of biotechnology research. Such funding levels are 1.7 times higher
than current funding levels for NSF grants and 2.8
times higher than those for USDA competitive grants
(NASULGC 1983). Starting salaries are much higher for
scientists trained in the new biotechnology due to
their relative scarcity and the higher demand for them
in the private sector. Indeed a survey of agricultural
biotechnology firms indicated a projected national need
of 434 new Ph.D.s between 1985-1987. In the short run,
these factors have led to a drain of scientific manpower from the universities to industry. Currently one
quarter of industry molecular biologists were previously employed as university faculty. A concentration
of scientific talent might also develop in a few
states. For example, six states report over twenty
FTEs devoted to biotechnology. They account for 42
percent of all SAES's FTEs for biotechnology but only
account for 28 percent of the total SAES's FTEs for
research (California, Michigan, Texas, Washington,
Wisconsin).

In the long run, however, the situation could
change considerably. First, the demand for new bio-
technologists will decline somewhat as some venture
capital firms go bankrupt. Second, the supply may
begin to increase as some universities are able to
mount needed training programs. Third, as there are no
restrictions on entry into biotechnology (as there are
on physicians and plumbers), new entrants' salaries
will likely decline. However, as the techniques and
procedures become more established and routine, instru-

mentation costs will decline, perhaps even faster than scientists' salaries, and lower paid technicians will likely replace some scientists.

The balance of short-term versus long-term research agendas, always a concern of the public sector, may be disrupted by the new biotechnology research and the private sector. As we discussed earlier, the private sector has short-term proprietary goals, and as a consequence, funding for research is also generally short-term. A 1984 study of biotechnology companies revealed that nearly half of all such firms fund research in universities. Of those funding research, 50 percent of the firms reported that projects last one year or less. Only 25 percent reported projects lasting more than two years. In contrast, 92 percent of NIH's extramural awards were for three years or longer (Blumenthal et al. 1986a; 1986b). Similarly, Hatch formula funding projects are generally of a long=term nature. Dependence on private sector funds will generally change not only the time frame but also the stability of funding and the substance of the research. Industry should not be able to direct researchers in the public sectorAs one private sector scientist noted, "The problem with university= industry biotech institutes is that they can compromise the universities. The establishment of private/public joint projects leads to the point where the public partner becomes the captive of the private sector." (personal interview 1984).

Perhaps more significant than the time frame for research, in regard to public sector agricultural research, is the increasing importance of the private sector's research agenda for public research. The scientific and commercial potential of biotechnology, as well as decreasing federal monies for research, have led to increasing university-industry ties. As mentioned earlier, the private sector provides 16-24 percent of all funds expended for biotechnological research in the SAES, while the national average of private funds for university research in general is 3-5 percent (NASULGC 1983, p. 12; Blumenthal et al. 1986a). In addition to higher funding levels, the nature of private sector grants has changed. Previously, fairly small grants went to individual scientists, and the grants came with few overt strings attached. For example, a seed company would give $10,000 to an SAES for vegetable breeding with the idea that any varieties developed or knowledge gained would be beneficial to the industry as a whole. Now, much larger grants are often given to a department or institution with specific conditions. The research, even though often of a basic nature, is viewed as proprietary. Patents or procedures developed from such research are, at the least, shared with the granting company. The need for the private sponsors to obtain a return on their investments must be balanced with the public universities' social contract. Many SAES are spending considerable time and money determining how to establish guidelines for university-industry research relationships. Indeed, SAES view royalties and licensing fees from patents and the legal protection of new plant varieties as a potential source of income for research support (NASULGC 1983; 1984; Blumenthal et al. 1986).

The potential money to be made has inhibited the flow of information among biotechnology scientists. This is particularly true with private sector grants. Scientists often must delay public discussion of work until it has been reviewed by the sponsoring company. In Blumenthal's study (1986a; 1986b), 25 percent of industrially supported biotechnology faculty reported that they have conducted research that belongs to the firm and cannot be published without prior consent while 40 percent of faculty with industrial support reported that their collaboration resulted in unreasonable delays in publishing. Even scientists with public funding feel inhibited about discussing their work for fear that a private company with the money, equipment, and time will perform experimental work before they can. The harm done by restricted scientific communication is not known (Buttel and Belsky 1987; Edwards 1987). However, many scientists in both the public and private sector view increasing secrecy as detrimental to science. The free flow of new information is fundamental to public sector science and, indeed, one industry scientist remarked that more knowledge is generated by keeping an open environment for scientists. Most breakthroughs don't come from just one lab; instead there is a need for information from a number of different labs. Communication among scientists is crucial (OTA 1986).

The investment of venture capital in the SAES, often tied to exclusive release of technology via patent rights or exclusive licensing, is viewed as an even more disturbing development than decreasing communication among scientists. Both public and private sector scientists have stressed the potential detrimental effects of granting private patents for work done in the public sector. Examples of potential detrimental effects include favoritism, unwarranted financial advantage through privileged use of information or technology partly derived from research using public funds, constraints on sharing of germplasm, and shelving of research that may be of interest to the public but not to the corporation. Also, scientists from both the public and the private sector wonder whether it represents a conflict of interest and believe that work done in the public sector should be available to the public. In addition, many believe it might unduly influence public sector research toward individual private sector goals and has the potential for changing the distribution of benefits from land grant research discoveries.

Finally, the costs of the resulting discoveries are internalized in the price of the resulting product. The price the public pays for the product also includes any monopoly rents associated with the conferral of the rights. Society thus pays twice: once for the costs of the research and then again for its benefits and products. "This establishes a dangerous precedent," explains one scientist. "It diverts researchers from unbiased basic research of benefit to the general public to biased research oriented to specific money making objectives. It also uses tax funded money and facilities for the benefit of the exclusive grantors" (personal interview 1984). Another industry scientist stated the issue somewhat differently when she said, "We need to be careful that research applications don't damage the public welfare, physically, environmentally, or aesthetically. A few privileged interest groups shouldn't be allowed to dominate the field" (personal interview 1984).

Such strong sentiments are not restricted to the scientists themselves. A number of research directors for seed companies stress that private sector grants should come with "no-strings attached." In addition, an ASTA survey of its membership revealed that a large majority of the seed companies responding were strongly against the use of venture capital for SAES research when it is tied to exclusive release (ASTA, 1984).

The new biotechnologies are also likely to have rather dramatic effects on farmers. For example, as the production of finished seed is moved more and more to the private seed industry, product differentiation of the type found in other consumer goods is likely to invade the farm sector. Farmers may be faced with a bewildering array of seed varieties. Thus, advertising is likely to play a much greater role in seed sales than it has in the past. In many cases, seeds come pelleted with chemicals. One particular concern is that it may be difficult for farmers to distinguish between product differentiation and significant varietal differences.

Another impact may be the continuation of the movement of farmers out of their traditional roles as the primary clients for plant breeding research. They are already being replaced by seed companies and the chemical companies that run them, as well as by processors. These large companies tend to have the money needed for biotechnology research and can influence research directions. For example, food processors are interested in the breakthroughs and proprietary advantages made possible through the new biotechnologies. Most biotechnology companies working with tomatoes are trying to increase total solids. Increasing total solids decreases water content thereby greatly increasing case yields, i.e., the amount of canned product per unit weight of harvested tomatoes, but not necessarily crop yield. An increase in total solids of 1 percent is worth an estimated $75 million annually (Davenport, 1981). Since processors either pay no, or a very small, bonus to growers for increased solids tomatoes, the processors gain the vast majority of benefits from a higher solids tomato. Furthermore, canners will only buy those tomato varieties that appear on their processing industry's acceptable list, thereby heavily influencing which varieties the farmers grow.

Scientists and administrators appear unaware of the potential for conflict between the interests of farmers and those of agribusiness. This potential for farmer-industry conflict was manifested in recent Senate testimony given on behalf of the American Farm Bureau Federation. The Farm Bureau argued that public research should be focused on reducing the cost of inputs, particularly chemicals (Hawley 1984). It is conceivable that the new biotechnologies could achieve that goal. Much of the evidence, however, suggests that just the reverse is likely. An experiment station director illustrated this conflict nicely when he told a group of state agribusiness leaders that "in the distant future I can foresee a perennial corn crop which fixes nitrogen, performs photosynthesis more efficiently, and is weed and pest resistant." With that one statement he managed to alienate every sector of agribusiness community.

A final consideration in the development of biotechnology research is the general funding for public sector research. Given the large infusion of private sector funds, the question is being asked whether the levels and type of industrial support is sufficient to justify reductions in federal support of biotechnology research at universities. The rich flow of venture capital into biotechnology may mean the government no longer need support that element of biomedical research so heavily. A similar argument was made recently in conjunction with public sector agricultural research by a number of senior congressional agricultural committee staff personnel attending a conference on agricultural research policy.

However, this raises a number of important questions. Blumenthal et al. (1986a) point out that industry funding remains small compared to government support of biotechnology research on the nation's campuses. Consequently, government funding remains an important cornerstone of academic research. Indeed a recent GAO (1986) report on university funding (1986) indicated that 60 percent of university research funding comes from federal agencies. If the federal government reduced its funding of biotechnology research by approximately 10 percent ($60 million), industry would have to increase its support by roughly 35 percent to compensate. Blumenthal et al. (1986a) conclude that this is unlikely to occur in the near future, if at all.

Secondly, even if industry increased its support to compensate fully for federal or state cutbacks, society might not be completely satisfied with the research agenda or results. As noted earlier, the private sector research agenda is shorter in term and more narrow in scope. Indeed, the new biotechnologies and the increasing concentration of research for proprietary products by both the public and private sector suggest the need for greater public funding for an array of research agendas that will not be addressed by these new research thrusts. In addition to the traditional non-proprietary mission-oriented research of the land grant system, the new biotechnologies may require an expanded role for research management as well as for technical environment and social impact assessment. These include such issues as risk assessment, genetic vulnerability, social equity, etc. If research policy makers do not want the land-grant universities to confer property rights to particular private corporations or to become extensions of the private sector, they must provide the level of funding necessary for competition with other non-land grant universities that confer such rights. As the recent OTA report on <u>Technology, Public Policy and the Changing Structure of American Agriculture</u> observed, "This funding decision is a basic public policy decision -- maybe the most basic decision since the land grant system was created (OTA 1986:276).

NOTES

1. This is a revised version of a paper presented at the meeting of the American Association for the Advancement of Science May 1986. The material in this chapter is based upon work supported by the National Science Foundation and the National Endowment for the Humanities under grant No. RII-8217306. Any opinions, findings, conclusions, or recommendations expressed in this publication are those of the authors and do not necessarily reflect the views of these organizations.

REFERENCES

American Seed Trade Association (ASTA). 1984. <u>Report of public research advisory committee</u>. ASTA.

Andrews, F. M., ed. 1979. <u>Scientific productivity: The effectiveness of research groups in six countries</u>. Cambridge: Cambridge University Press.

Blumenthal, D. M.; Gluck, M.; Louis, K.; and Wise, D. 1986a. Industrial support of university research in biotechnology. <u>Science</u> 231:242-46.

Blumenthal, D.; Gluck, M.; Louis, K. S.; Stoto, M. A.; and Wise, D. 1986b. University-industry research relations in biotechnology: Implications for the university. <u>Science</u> 232:1361-66.

Brill, W. J. 1985. Safety concerns and genetic engineering in agriculture. Science 227:381-84.

Burkhardt, J.; Busch, L.; Lacy, W.; and Hansen, M. 1986. Biotechnology and food: a social appraisal. In Food biotechnology, ed. D. Knorr, pp. 575-600. New York: Marcel Dekker, Inc.

Busch, L., and Lacy, W. B. 1983. Science, agriculture, and the politics of research. Boulder: Westview Press.

Busch, L., and Lacy, W. B. eds. 1986. The agricultural scientific enterprise: a system in transition. Boulder: Westview Press.

Buttel, F. H.; Kenney, M; Kloppenburg, J. Jr.; Smith, D; and Cowan, J.T. 1986. Industry/land grant university relationships in transition. In The agricultural scientific enterprise: a system in transition, eds. L. Busch and W. B. Lacy, pp. 296-312. Boulder: Westview Press.

Buttel, F.H., and Belsky, J. 1987. Biotechnology, plant breeding and intellectual property: social and ethical dimensions. Science Technology, and Human Values 12(1):31-49.

Cochrane, W. W. 1979. The development of American agriculture: a historical analysis. Minneapolis: University of Minnesota Press.

Davenport, C. 1981. Sowing the seeds-research, development flourish at DeKalb, Pioneer Hi-breed. Barron's 61(9): 9-10, 33.

Diamond vs. Chakrabarty. 1980. U.S. Reports 477:303-322.

Doyle, J. 1985. Altered harvest: Agriculture, genetics, and the fate of the world's food supply. New York: Viking.

Edwards, I. 1987. Biotech industry should consider impact of plant patents on future of crop improvement. Genetic Engineering News 7(8):4,31.

Fischhoff, D. A.; Bowdesh, K.S.; Perlak, F. J., et al. 1987. Insect tolerant transgenic tomato plants. Bio/Technology 5:807-13.

Freeman, K. 1987. Biotech research/industry parks a boon for universities and businesses. Genetic Engineering News 7(8):12,13.

General Accounting Office (U.S.). 1985. Biotechnology: the U.S. department of agriculture's biotechnology research efforts. GAO/RCED-86-39 BR. Washington, D.C.: U.S. Government Printing Office.

_____. 1986. University funding: assessing federal

48

funding mechanisms for university research.
GAO/RCED-86-75. Washington, D.C.: U.S. Government
Printing Office.
Genetic Engineering News. 1985. Fourth annual GEN
guide to biotechnology companies. Genetic Engineering
News 5(10):24-47.
_____. 1987. Sixth annual GEW guide to biotechnology
companies. Genetic Engineering News 7(10):21-54.
Giedion, S. 1975. Mechanization takes command. New
York: W. W. Norton.
Glazer, S. 1986. Businesses take root in university
parks. High Technology (Jan):42-47.
Hansen, M.; Busch, L.; Burkhardt, J.; Lacy, W. B.;
and Lacy, L. R. 1986. Plant breeding and biotechnology: new technologies raise important social questions. BioScience 36(1):29-39.
Hawley, B. 1984. Statement of the American Farm Bureau
Federation to the Subcommittee on Agricultural Research and General Legislation of the Senate Agriculture, Nutrition and Forestry Committee, 14 June.
Jaschik, S. 1986. Universities' high-technology pacts
with industry are marred by politics, poor planning and hype. The Chronicle of Higher Education
XXXI (March 12:15-18).
Kalter, R. J. et al. 1984. Biotechnology and the
dairy industry: production costs and commercial po-
tential of the bovine growth hormone. Ithaca:
Cornell Department of Agricultural Economics.
Kalton, R. 1984. Who is going to breed the minor crops
and train the next generation of plant breeders. Seed
World 122(10):16-17.
Kenney, M. 1986. Biotechnology: The university-indus-
trial complex. New Haven: Yale University Press.
Kloppenburg, J. 1987. First the seed: the political
economy of plant biotechnology, 1492-2000. New
York: Cambridge University Press.
Krimsky, S. 1984. Corporate academic ties in biotech-
nology: a report on research in progress. Gene
Watch (Sept./Dec.):3-5.
Lacy, W. B. and Busch, L. 1983. Informal scientific
communication in the agricultural sciences. Informa-
tion Processing and Management 19(4):193-202.
Lacy, W. B., and Busch, L. 1984. The role of agricultural research for U.S. food security. In Food security in the United States, eds. W. Lacy and L. Busch,
pp. 289-320. Boulder: Westview Press.
Makulowich, J. 1987. Biotech's growth depletes aca-

demic ranks. <u>Genetic Engineering News</u> 7(9):1,28.

Matthiessen, C., and Kohn, H. 1984. In search of the perfect tomato. <u>The Nation</u> July 7-14.

Mehta, V. 1987. Investment potential increases in Europe as biotechnology industry slowly matures. <u>Genetic Engineering News</u> 7(7):4,9.

Merges, R. 1987. Assessing the impact of higher life form patents on the biotech industry. <u>Genetic Engineering News</u> 7(6):24,25.

Miller, L. I. 1985. Biotechnology mergers signal industry consolidation. <u>Genetic Engineering News</u> 5(2).

Mooney, P. R. 1980. <u>Seeds of the earth</u>. Canadian Council for International Cooperation and the International Coalition for Development Action.   Ottawa: Inter Pares.

_____. 1983. The law of the seed: another development in plant genetic resources. <u>Development Dialogue</u> 1-2:104.

National Academy of Sciences. 1984. <u>Genetic engineering in plants</u>. Washington, D.C.: NAS.

National Association of State Universities and Land Grant Colleges. 1983. <u>Emerging biotechnologies in agriculture: issues and policies. Progress report II</u>.

_____. 1984. <u>Emerging biotechnologies in agriculture: issues and policies. Progress report II</u>.

_____. 1985. <u>Emerging biotechnologies in agriculture: issues and policies. Progress report IV</u>.

National Research Council (U.S.), 1987. <u>Agricultural biotechnology: Strategies for national competitiveness</u>. Board on Agriculture, Committee on a National Strategy for Biotechnology in Agriculture. Washington, D.C.: National Academy Press.

Netzer, W. 1987a. Researchers seek ways to raise crop yields by improving photosynthesis. <u>Genetic Engineering News</u> 7(10):8, 70.

_____. 1987b. Phosphate solubilizing genes might revolutionize fertilizer technology. <u>Genetic Engineering News</u> 7(8):10, 41.

Office of Technology Assessment (OTA), U.S. Congress. 1984. <u>Commercial biotechnology: an international analysis</u>. OTA-BA-218. Washington, DC.: U.S. Government Printing Office.

_____. 1986. <u>Technology, public policy, and the changing structure of American agriculture</u>. OTA-F-285 Washington, D.C.: U.S. Government Printing Office.

Pelz, D. C., and Andrews, F. M. 1976. _Scientists in organizations_. Ann Arbor: Institute for Social Research.

Rogoff, M., and Rawlins, S. L. 1986. _Food security: a technological alternative_. Unpublished paper.

Rose, C.M. 1987. _Agricultural research: issues for the 1980's_. Congressional Research Service Report for Congress 87-430 SPR, Washington, D.C.

Schrage, M., and Henderson, N. 1984. Biotech becomes a global priority. _Washington Post_ Dec. 17

Sohn-Rethel, A. 1978. _Intellectual and manual labor_. Atlantic Highlands: Humanities Press.

Solomon, T., and Tornatzky, L.G. 1986. Rethinking the federal government's role in technology innovation. In _Technology innovations: Strategies for a new partnership_, eds. D. O. Gray, T. Solomon, and W. Hetzner, pp. 31-48. Amsterdam: Elsevier Science Publishers.

USDA-ARS. 1983. _Agricultural research service program plan, 6 year implementation plan, 1984-1990_. Washington, D. C.: Department of Agriculture.

Zuiches, J. J. 1986. Research funding and priority setting in state agricultural experiment stations. In _The agricultural scientific enterprise: a system in transition_, eds. L. Busch and W. B. Lacy. pp. 97-110. Boulder: Westview Press.

Wheeler, D. L. 1988. Harvard University. _The Chronicle of Higher Education._ receives first U.S. patent issued on animals XXXIV (32):1

3. Regulation of Agricultural Biotechnology: Historical Perspectives

The evolving regulations for agricultural biotechnologies can have important implications for the types of products that will be developed and the speed with which they reach commercialization (McGaritt 1985; Alexander 1985; Hardy and Glass 1985). Products that release genetically engineered organisms into the environment receive heavy scrutiny from regulatory agencies, and their dates of market entry will most likely be significantly delayed.

The current regulatory environment concerning biotechnologies has the potential for significant shifts in policy due to the speed with which these technologies become available and the lack of scientific basis for judgments on risk. An attempt is currently underway to coordinate the regulatory activities of the Environmental Protection Agency (EPA), the Food and Drug Administration (FDA), and the U.S. Department of Agriculture (USDA) so that biotechnology is adequately covered by existing laws.

Several areas of caution have been suggested (Kenny 1985; Conservation Foundation 1985). Critics of the regulatory process have charged that little research has been completed to assist in developing appropriate regulatory guidelines. More coordination is needed among the several agencies where regulatory jurisdiction may overlap. Some critics argue that the EPA's budget is insufficient to handle the normal flow of new products. If a flood of new biotechology products materializes as expected, the agency might not be able to provide effective control (McGaritt 1985). Long-term effects of biotechnologies are not known and are not likely to be known when regulatory decisions

are needed. Regulation may be overwhelmed by the
sheer momentum of public and private R&D efforts (Hardy
and Glass 1985).

Insights can be gained by examining the history
of safety regulation in agriculture to anticipate the
evolving regulatory regime for agricultural
biotechnologies. Biotechnology related release of new
organisms into the environment has parallels with the
introduction of exotic pests. Biotechnologies that
facilitate production of a naturally occurring
substance involve scientific processes and potential
risks analagous to those experienced in the development
of pesticides. Thus, the history of regulatory policy
for pesticides and exotic pests will be used as a guide
for the regulatory policy that might be expected for
agricultural biotechnologies. This chapter will
proceed by providing historical background on the
regulation of pesticides and exotic pests and by using
this historical perspective to draw precedents for
future regulation of agricultural biotechnology.[1]

HISTORICAL PERSPECTIVE ON
SAFETY REGULATION IN AGRICULTURE

Pesticides

Public sector institutions, agricultural producers, technical change, and the scientific community
were all important elements in the historical evolution
of pesticide regulation. By the turn of the century,
several insect epidemics had precipitated the need for
improved methods of insect control. The first widely
used chemical insecticide was Paris green (copper
aceto-arsenate) in the 1870s (Dunlap 1978). In the
1890s the boll weevil crisis brought public attention
to farmers' inability to control certain pests. The
profession of economic entomology was born during this
era as were the Federal Bureau of Entomology and agricultural experiment stations.

Chemical insecticides, biological control, and
cultural practices are the major means for controlling
insects. From 1895 to 1918, entomologists advised
farmers to use cultural practices in the control of the
boll weevil, with little success. However, when cal-
cium arsenate was found to be effective against the

weevil, farmers responded by purchasing ten million
pounds in 1920. Biological control was lauded when in
1890 a vedalia beetle imported from Australia successfully controlled a scale in California. However,
the complete failure to control the gypsy moth in New
England in 1890 convinced entomologists that much more
knowledge of the insect and its behavior was necessary
to make biological control viable. Consequently, by
the 1920s chemical insecticides had become the favored
tool. Cultural control caused social and political
disruptions, and biological control required more
sophisticated techniques. But chemicals were convenient, inexpensive, and readily available (Dunlap
1978). DDT and chlordane, introduced in the 1940s,
appeared to be ideal chemicals.

The scientific community was not unanimous in its
concern over the potential hazards of increased chemical insecticide use. The conflict was centered around
the type of experiment undertaken to determine the
safety of a chemical. The FDA took the approach that
chronic effects must be examined and that such effects
were best studied using experimental animals exposed
over a lifetime (Dunlap 1978). The Public Health
Service downplayed the importance of chronic effects
and instead conducted population surveys of heavily exposed areas to obtain data on residues in body tissues,
characteristic symptoms of illness, or susceptibility
to disease. Economic entomologists and the
agricultural producers tended to support the latter
approach, while medical groups were more likely to
favor the FDA approach.

In 1935 the "lead arsenate crisis" prompted the
American Medical Association to assert the gravity of
the risk to public health from insecticide residues and
to recommend stricter regulation of their use (American
Medical Association 1935). This debate culminated in
the enactment of the Federal Food, Drug, and Cosmetic
Act in 1938. Dunlap summarizes the regulatory climate
during the era of the development of DDT, chlordane,
and the other chlorinated hydrocarbons:

> That the fight was one-sided is clear evidence of
> the political power of the growers and of the
> public's lack of interest in the problem. The
> revision of the food and drug laws in the 1930s,
> during which the two factions in the residue
> battle attempted to have their ideas written into

the law, attracted little interest. David Cavers, a law professor who, in 1933, helped draft the first version of what became the 1938 act, wrote in 1939 that "perhaps the most striking characteristic of the history of the Food, Drug, and Cosmetic Act is the fact that this measure ... never became the object of widespread public attention, much less of informed public support."

By the time DDT became available for civilian agricultural and public health use in 1946, both the Food and Drug Administration and the Department of Agriculture were accustomed to acting as service agencies for interest groups. Their concern, except where there was an immediate danger to public health or safety, was to work with the affected industry and to enforce compliance without causing the industry more difficulty than necessary -- the industry being the judge of the hardship. Whether, under the circumstances, they could have followed different policies is another question, but given their need for annual funding, the lack of broad public support for a strict policy, and the imprecise nature of the threat to public health, it is difficult to see how else they could have acted. From the viewpoint of the 1970s the policy may seem less than wise, but this is largely because of our increased knowledge (Dunlap 1981, p. 55).

The regulatory climate of the 1970s differed immensely from the climate of the 1940s according to Dunlap. The public was brought into the debate in the 1960s and its presence was reflected in the creation of the EPA (Marcus 1980). The present more restrictive regulation follows the FDA approach where chronic effects are examined. The widespread use of persistent chlorinated hydrocarbons is in part responsible for public concern over residues and the resultant strict environmental policy.

The Federal Insecticide, Fungicide, and Rodenticide Act (FIFRA) of 1947 controlled pesticides marketed in interstate commerce. Administered by the USDA, this act required that all pesticide products distributed in interstate commerce be safe and effective. In addition, the FDA prohibited feed or food residues exceeding an established safe tolerance level. To obtain registration of a new pesticide, a producer had to demonstrate its safety.

FIFRA exerted its principal influence on research and development activity in three ways: (1) registration and labeling, (2) experimental use permit, and (3) establishment of tolerance. Toxicity and efficacy tests, supporting data, and labeling were required for pesticide registration. The nature and proposed use of the pesticide determined the stringency of toxicological data required.

Tests performed under field or laboratory conditions were designed to demonstrate the product's ability to control the specified pest (following the label directions) with no injurious effects upon the crop or land. Applications for registration were reviewed by other agencies to determine the impacts on wildlife (the Department of Interior) and human safety (the Department of Health, Education and Welfare). In the case of a product with potential residues, the USDA sent the application to FDA for a tolerance or exemption decision.

The experimental use of pesticides was not explicitly addressed by the original FIFRA legislation; the temporary permit program was enacted through amendments in 1964. Experimental use permits were not difficult to obtain. Much research and development activity was undertaken through agricultural extension programs at state universities, where no permits were required (Wechsler et al. 1975).

Research and development activity in the pesticide industry was also affected by the Delaney Amendment. This revision of the Food, Drug, and Cosmetics Act in 1962 prohibited any measurable quantity of any known carcinogenic pesticide or its residue on food.

Essentially as a result of improvements in scientific measurement, further amendments were made to FIFRA in 1967. Previously, registration could be obtained on a "negligible residue" or "zero tolerance" basis under Section 346a of the Federal Food, Drug, and Cosmetics Act. A reevaluation and reregistration program was instituted that required a finite tolerance.

The Federal Pesticide Act of 1978 contained several significant changes in pesticide policy. In general, the revisions were intended to ease the testing requirements with a minimum adverse impact on the environment. Major features involved data compensation, state authority, permissible label inconsistencies, conditional registration, minor use data requirements, efficacy data requirements, and generic approach to registration. Concerns that the regulatory process had become excessively restrictive prompted the attempt to "...streamline the pesticide regulatory process, encourage pesticide research and development, and provide greater pesticide use flexibility for growers" (CAST 1981, p. 21).

Concern has been expressed over the disincentive to undertake research and development of minor use pesticides and the resultant concentration of R&D effort on large potential markets (CAST 1977). Federal and state governments have established programs (IR-4) to facilitate minor uses. Consequently, data requirements can be eased in consideration of volume, pattern and extent of use, cost of needed data, and degree of exposure to humans and the environment. "Waiving efficiency requirements under certain conditions, increasing state authority to register pesticides, easing labeling requirements, and allowing conditional registration should help alleviate the minor use pesticide problem," according to a report by the Council for Agricultural Science and Technology (CAST 1981, p. 19). Additional features include the possible waiver of efficiency data for any material and the move toward the generic approach to registration.

Exotic Pests

The problem posed by free release biotechnology has been characterized as posing "low probability/ high consequence" risk (Vanderbergh 1986). This characterization of risk is exactly what is posed by the introduction of an exotic pest. Although most such introductions fail (in the sense that the introduced species proves it can not survive in the new environment), there have been notable exceptions, including the gypsy moth, starling, and kudzu vine in the United States. The chestnut blight all but eliminated the American chestnut tree from the Eastern United States (Sill 1982). The Hessian fly was estimated to have caused up to a 50 percent yield loss in wheat until successful control measures were developed (Schwartz and Klassen 1981). For many years, the European rabbit was a major crop pest in Australia. Against these potential problems, we can note that there is a small probability that any particular organism can upset an ecological system. For example, the Animal and Plant

Health Inspection Service (APHIS) of the USDA reported intercepting 189,907 plant pests in 1986. APHIS does not claim to be a perfect filter, so it is clear that numerous accidental introductions of potential plant pests simply are undetected because the introduced species do not survive. Simberloff examined reports of introduced species, finding that 678 of 854 failed to have any impact on the resident species. Only 71 documented extinctions of resident species were found. The sample is admittedly biased since introductions exhibiting a dramatic effect are more likely to be reported.

We define a pest as any organism that is detrimental to humans and their activities. Exotic pests are defined as those pests that do not currently live in a given area (defined by either geographical or political boundaries). If such pests were to be introduced into a new area, they could cause economic losses.

Only a limited number of mechanisms exist to prevent introduction of unwanted organisms. National quarantines have been in effect since 1912. "The primary authority provided by the Plant Quarantine Act of 1912 is a means to control the artificial introduction of exotic plant pests associated with plants" (Rohwer 1985, p. 253). Since then, other legislation has further strengthened and expanded the use of quarantines. Even state-level quarantines are not unusual.

Evaluating these pest control methods is problematical if pest damage has not occurred. It is usually hard to determine whether a method has worked or if other uncontrolled forces have removed the pest. Of course, there are a few cases where one can evaluate the employed technique. Quarantines were successful in preventing reentry of the black potato wart (Sill 1982). But a perfectly successful program would never show damage from the pest, making the effect of the program unmeasurable.

Once an exotic pest has become problematic there are several control measures. Areawide eradication is one option. The most recent publicized example of areawide eradication is the California program against the Mediterranean fruit fly (Conway 1982). Most of the work to control the pest occurred in the summer of 1982. Areawide programs can take many forms. Both large-scale chemical applications and the release of the pest's parasites and predators have been employed.

For some insect pests, sterile males can be released to disrupt mating and reproduction. Mandatory cultural practices can sometimes be initiated.

Plant breeding for resistance is another means of coping with exotic pests. Plant breeding can be undertaken before or after entry. Although plant breeding has occasionally been successful, it is a slower control measure than the chemical method. The attempt to control the Hessian fly is an example of plant breeding after entry. This fly was supposedly introduced during the American Revolution in the bedding of the Hessian troops (Pfadt 1971). Breeding wheat varieties for fly resistance began in the 1930s and still continues. These resistant varieties, in conjunction with altered planting dates, have reduced the Hessian fly's effect to only minimal economic loss.

Few people are satisfied with any of these solutions, or attempts to mitigate the effects of exotic pest damage. These actions raise questions of expense, safety, and government control. There is no guarantee that such measures will be successful at removing the pest or recreating environmental characteristics that existed before pest damage occurred. In popularized discussions of deliberate release of genetically engineered organisms, industry critics have argued that potential damage is large compared to introducing more familiar organisms (Doyle 1985). Government regulators are likely to be even less certain about the impacts of open-air testing and research than about ordinary exotic pests.

REGULATION OF AGRICULTURAL BIOTECHNOLOGY

Non-release Biotechnology

The admittedly short history of genetic engineering regulation indicates that products receive different regulatory treatment given the answers to two key questions: (1) Does the product involve release into the environment of organisms where that organism is intended to survive, at least for a time? and (2) Will the new product become part of the human food supply? So far, those products for which the answer to the first question has been affirmative (for example, "ice minus") have been treated as if the microorganism were an exotic pest. Those products for which the second question is germane have received a regulatory treatment that shares some characteristics with agricultural chemical regulation.

These two key questions for new biotechnology product registration parallel questions that have been faced by prospective agricultural chemical products: (1) Is the new product carcinogenic? and (2) Does the registration involve a food use? Just as the possibility of carcinogenicity for new chemicals elicits a debate on "zero risk," potential uncontrolled release into the environment provokes strong debate over "acceptable risk."

The "food use" and "food supply" questions differ in one important aspect. In general, there was little debate over whether an agricultural chemical involved a food use. With new biotechnology products, more debate is likely over whether the product enters the human food supply. The public has shown minimal concern over the various hormones that are present in the meat supply. Initial indicators point to the possibility that consumers will be more concerned about recombinant DNA-derived hormones that might be used in commercial livestock production than they have been regarding the current technology (Jones, 1986). Jones suggests that the key animal biotechology regulatory issues can be grouped into process and product. The regulatory process for animal research using genetic engineering will closely follow animal drug or animal biological procedures. The products of animal biotechology research will face questions about safety and identity. A precedent for inspection of the quality of new products has been set from experience with interspecific hybrids of cattle and buffalo.

> ...the inspection of cattle/buffalo hybrids has established a phenotypic (based on appearance) criterion for deciding: how novel food animals should be inspected. As the genetic engineering of food animals on a production basis draws nearer, it may be necessary to supplement the phenotypic criterion with genetic (based on pedigree) criterion to assure that the essential characteristics of animals slaughtered under current food statutes are maintained (Jones 1986, p. 281-2).

Jones (1985) has developed a regulatory model for gene transfer in food organisms. The model includes three phases: delivery, expression, and commercial. Delivery is completed when the gene is incorporated into the host organism. Expression occurs with the "achievement of controlled expression of the transplanted genetic material in the host organism." An advantage of the model is its flexibility to adjust to the on-going public safety debate in the courts and Congress.

Bovine somatotropin (BST) and other products that do not involve environmental release raise an entirely different set of questions. Unlike free-release agricultural biotechnologies, BST does not require deliberate release of a genetically engineered organism. The development of a process for its industrial manufacture solved the problem of producing large quantities of the substance inexpensively. By injecting BST, dairy farmers can, theoretically, augment the BST that dairy cows normally produce. If administered at the correct point in the lactation cycle, BST could markedly increase milk production. Because BST is a naturally occurring substance, its use and release into the environment has not raised questions of changing ecological processes by introducing a new organism. The risks considered by regulatory agencies include the potential for damage to dairy cow health and the question as to whether milk products contain substances harmful to consumers. Unlike the deliberate release case, if these risks do exist, controlled use of BST could probably limit those effects.

Free-Release Biotechnology

The deliberate release of gene-altered organisms raises many of the same problems as use of toxic chemicals: how chemicals are changed and where they will be moved; hazards; exposure; and effects on ecosystems (Levin 1986; Levin and Harwell 1985). The key regulatory issues for the environment center on rules of containment, which specify testing procedures that ensure that the microorganism will not enter the ecosystem. There has always been a special exemption from strict containment requirements for research and development; however, no such exemption is foreseen for biotechnology.

Levin and Harwell (1985) note the possibility of uncontrolled reproduction of some genetically engineered organisms makes the problem resemble introducing an exotic pest. However, many scientists argue that there is a significant difference between introducing a new species vis-a-vis an old species with an altered gene. This poses a problem for regulatory agencies because the outcome of any such release is highly uncertain.

Regulators' uncertainty is the result of not fully understanding how an organism and its environment interact. A pest may be of little or no economic importance in its native land due to a natural system of checks and balances. However, introducing the pest into an area with a favorable climate and susceptible host, may increase the pest population dramatically, causing significant losses (McGregor 1973). Likewise, an apparently suitable environment in a new area may unexpectedly not support pest or pathogen development, adding to regulatory uncertainty.

In either case, creating a novel genotype and releasing it into the open environment, or physically moving an existing species from one set of environmental conditions to another, means that one cannot be certain of the organism's population dynamics in the new environment. Data on the behavior of an organism can be conjectured from field observations in areas where it is endemic and from controlled-condition greenhouse and growth chamber experiments (e.g., Bromfield 1980). However, whether these experiments and observations are carried out with more or less constraints on population growth can only be answered, with certainty, by releasing the organism. Hence, any forecast entails some degree of speculation on whether the release of a novel genotype would be either harmless or harmful

Regulatory agencies face pressures not just for protection from environmental problems, but also for allowing new products onto the market. If the only questions genetic engineering applications raised were safety, agencies could meet such goals by forbidding research, testing, development, and marketing of genetically engineered products. However, the companies that invest in research and development make demands on regulatory agencies to allow their products to reach the market. Also, regulatory delays impose a real cost on society because food and fiber production is lower than it might have been in production carried out with innovative inputs. Delaying introduction of new products could have important implications for international competitiveness.

For example, Biologics Corporation, manufacturer of a vaccine for the swine virus pseudorabies, has won, lost, and regained its USDA registration to sell the product. The vaccine is created through gene deletion, making it a weakened but live form of the pseudorabies virus. Charging that USDA had not followed its own guidelines for product assessment, the Foundation on Economic Trends, headed by Jeremy Rifkin, sued. The vaccine's registration was canceled although subsequently re-issued. While the Foundation's complaint was filed on the basis of improper regulatory procedure, the real issue centered on the potential hazards from deliberate release of gene-altered organisms. In re-issuing the license, USDA argued that the deletion of the single gene prevented the virus from producing an enzyme it requires to multiply and spread.

Frostban is a bacteria designed to inhibit the formation of frost on plants. This gene-deleted product is identical to the naturally-occurring bacteria that inhabits plants except that it is missing that part of the genetic code that triggers production of an ice-nucleating protein promoting frost.

In 1984, a University of California test of Frostban was prevented by a court ruling that the National Institutes of Health (NIH) had illegally given its approval. In late 1985, a private company, Advanced Genetic Sciences, obtained an experimental use permit from the EPA. The EPA then revoked the permit in a dispute over the company's testing procedures. In August 1986, University of California scientists again proposed a test of the bacteria, but a temporary restraining order delayed the test while an environmental impact statement was prepared. Open-air testing was delayed three years. Again, evaluation of the potential risks of releasing a novel genotype was at the crux of the debate.

Many public and private research organizations have attempted to transfer the insecticidal properties of <u>Bacillus thuringiensis</u> Burliner (commonly known as Bt), to other organisms. Monsanto was one of the first, developing a soil microbe altered to carry the toxins. The engineered microbe was intended to control soil-dwelling insects, reducing the need for synthetic organic chemicals.   The EPA postponed field testing in 1986.

Decision-Rules for Free-Release Organisms

The important questions about introduced altered organisms relate to disruption of ecological processes and potential health hazards.   Before these questions are asked, one has to know whether the new organisms will survive, and if so, whether they will reproduce. This is the type of speculation free-release biotechnology regulation requires.   To answer these questions and their implications, biologists have classified exotic species into four groups: (1) slightly modified forms of resident types; (2) forms that exist naturally in the target environment, but that require continual supplemental support or continual replacement to be sustained; (3) forms that exist naturally elsewhere, but that have not previously reached the target environment; and (4) genuine novelties (Levin and Harwell 1985, p. 59).   This classification scheme is useful in analyzing the deliberate or accidental release of genetically engineered organisms because it highlights the problems in forecasting externalities.   A generally accepted classification scheme could lead to some products receiving less regulatory scrutiny.   This classification scheme, or some very similar ranking appears to be behind some recent regulatory decisions.

The three genetically engineered agricultural products that involve release of an organism may be categorized by the above scheme despite little or no actual experience with two of them.   Frostban is identical to naturally occurring bacteria, except for a deleted gene.   Because Frostban is produced by gene deletion, the altered organism might be weakened and might need continual support or replacement, as described in the second category.   Gene deletion renders the bacteria incapable of invading plants through frost damage.   Hence, Frostban might be unable to compete with natural bacteria in the long run.   The pseudorabies vaccine, also produced through gene deletion, appears to fall into the second category. The weakened form of the virus makes it suitable as a vaccine.   The Monsanto soil toxin, although composed of well understood organisms, has been considered a genuine novelty, as described by the last category.

In June 1986, the White House issued a set of

comprehensive rules for regulating biotechnologies. These rules both coordinate the activities of relevant government agencies and attempt to answer questions regarding expectations about the survivability of gene-altered organisms.

The new rules entail only limited regulation of the products of gene deletion. Some biologists argue that gene deletion renders an organism harmless because some of the survival tools with which it evolved have been taken away. Biologists, however, are not unanimous in their support of the new rules. Some have argued that DNA exists in a delicate balance, and that disruption by gene deletion could cause an organism to mutate and multiply with unforeseen consequences (Schneider 1986). Regal argues that classifying genetically engineered organisms as harmless is based on simplistic ideas and will lead to a false sense of security. The judgment that an organism can or cannot find an ecological niche for itself depends on the speculator's personal belief in the development and stability of ecological systems. Regal counters the rosy scenario that ecological inertia might eliminate potential hazards by arguing that the gene transfer experiments now being done far exceed what can be done through ordinary breeding programs or sexual recombination. That is, laboratory-generated genetic variability may be greater than what occurs naturally.

Under the new regulations, fewer industry and federal agency resources would be required in the evaluation of products from biotechnology. Responsibility for oversight of federally-funded genetic engineering research originally rested with a committee operating under the auspices of NIH. As the field of inquiry broadened beyond medicine, other agencies became involved through re-interpretation of existing legislation on regulatory responsibilities. As currently administered, NIH will continue to monitor research activities, while responsibilities for oversight of specific applications will rest with other agencies. The USDA will regulate gene-altered animal vaccines and plants. While the EPA has general responsibility for oversight of genetically-engineered microbes, it will share duties with the USDA when agricultural crops are involved. Animal drugs and human health care products are the responsibility of the FDA. The Occupational Health and Safety Administration (OSHA) will look after genetically-engineered products used in the workplace.

Rule-Making Difficulties

The attempt to classify some organisms and types of manipulations as benign has not yet been successful. Discussions within the EPA have shown that it is not physically possible to test an organism's response to all the various environmental conditions it might encounter. Without such conclusive evidence, one can not make statements with absolute certainty about pathogenicity (Pesticide & Toxic Chemical News 1987). This ambiguity raises an impossible question for the biotechnology industry. Opponents of genetic engineering argue that more questions need to be answered before research is conducted (Rifkin 1984). With a seemingly infinite supply of troubling questions, biotechnology opponents could indefinitely stall the industry. Therefore, the real intention of demands for additional information cannot always be known.

The current regulatory apparatus for biotechnology is "a patchwork system of many agencies regulating products, applications, and experiments under preexisting and often outdated laws" (Jaffe 1987, p. 547). "The EPA's attempts to graft biotechnology onto a statute primarily designed to control inert pollutants will result in delaying litigation, prolonging the uncertainty among biotechnology enterprises" (Vandenbergh 1986, p. 1567). Specifically, Jaffe (1987, p. 547-8) argues that the regulatory system for biotechnology has many shortcomings that need immediate solutions.

The present system has many gaps, conflicts, and ambiguities. Who regulates deliberate release experiments done by industry? What biotechnology products can TSCA legally regulate? How do individual agencies assess the merits of deliberate release? Do the NIH Guidelines really regulate anything? Can an agency regulate biotechnology and promote it at the same time? Is there any overall federal interagency coordination?

The Vandenbergh critique of the current regulatory system for biotechnology focuses on the need for the development of a central databank of collective government agency experience with the assessment of the risks of the deliberate release of recombinant DNA-derived organisms. The organization charged with the coordinating function of biotechnology regulation, Biotechnology Science Coordinating Committee, could be strengthened by implementing such a databank. Like Jaffe, the Vandenbergh analysis pointed to the need for Congress to clarify TSCA's role in biotechnology regulation. In addition, research scientists need to be allowed to play a greater role in regulating these new technologies.

CONCLUSIONS

The federal regulatory system is designed to evaluate the environmental and human health risk of the products of new biotechnologies. The White House rules reflect this orientation in parceling out regulatory responsibilities, allowing such agencies as the EPA and the FDA jurisdiction over matters related to agriculture.

The current regulatory regime has been described by some critics as a "patchwork" with "many gaps, conflicts, and ambiguities," while others have been supportive of the decision not to establish a separate agency for biotechnology regulation. The concerns raised about use of these new products transcend safety issues; questions about their desirability involve impacts on the economic structure and social fabric of agriculture. Thompson (1987) argues that a fundamental constraint in developing effective regulation of biotechnology is the lack of public confidence in the scientific community whose arguments appear self-serving.

Biotechnology regulation is evolving now. The initial impasse, in which no field testing of genetically engineered organisms took place in the United States has been resolved. Several kinds of gene-deleted microbes have been released. Open-air testing has occurred both with EPA experimental use permits and without prior approval. Many in the scientific community predicted the absence of any deleterious effects from the legal initial releases.

The failure to observe any harm from a small number of experiments does not refute the low probability/high consequence risk concept often associated with free release. The exotic pest analogy leads one to expect no impacts from any particular organism. Whether that analogy will become a theoretical underpinning for genetic engineering regulation is still an unanswered question. An accumulation of new products tested and marketed without incident could lead some to conclude that such products are acceptably safe and, hence, require little regulatory scrutiny. Unfortunately, testing products without incident only gives regulators examples in which free release is harmless. The only way events can prove one of the competing theories is through an environmental disaster, implying that the exotic pest analogy is appropriate.

The regulatory decisions so far can be interpreted in a variety of ways. For example, the testing of several organisms that were produced through gene-deletion could suggest that that particular technology is considered acceptably safe and hence requires relatively little regulatory scrutiny. But again, these examples do not prove how regulatory agencies are likely to behave. Regulatory agencies may operate on a case-by-case basis, independently examining the merits of each product as it is presented.

Biotechnology industry representatives, recognizing that the industry will be regulated, have argued that existing laws and regulations provide sufficient safety standards for the new industry and that no new regulations or regulatory agencies are necessary (Gold-hammer 1986). They argue that recombinant DNA or gene-splicing techniques are not different from classical genetic selection. This argument is equivalent to demanding regulatory standards similar in form to those imposed on the pesticide industry. That is, the industry has demanded a finite set of well-defined goals that when successfully accomplished ensure the right to product commercialization. Eventually a social choice will be made: methods for assessing risk and levels of acceptable risk will be revealed. Such standards can be formal declarations from regulatory agencies or made, de facto, through the legal system.

The demand for regulatory standards is not a homogeneous force. The arguments over gene deletion and whether a product becomes part of the food supply indicate that some product developers believe their products should require relatively less regulatory scrutiny.

If the regulatory experience with agricultural chemicals and animal drugs is a good indication of the tendencies of regulatory bureacracy, attempts will be made to develop a classification scheme for new biotechnology products. Biological pesticides are regulated under a "tier system" whereby results on certain "screening" tests indicate the need for further testing. If safety is indicated in the first tier of tests, further testing will not be necessary. Although agricultural chemicals in general are not regulated under a formal tier system as used for biologicals, the concept of using certain test results as "triggers" for the need for further testing is common. Several such triggers have been discussed (e.g. free-release, human food supply, and exotic species). To the extent that such classification schemes can be developed, interaction between regulators and new product applicants will focus on where an individual new product fits into the scheme.

Arguments over BST reveal the importance product developers might place in having regulatory standards. The FDA statements indicate that the agency believes milk produced in current trials is safe for human consumption (Borcherding 1987). Opponents of the use of BST have focused on animal health and sociological issues (Annexstad 1986). The latter issue is new to agricultural technology regulation. Whether these newer types of questions might become part of imposed regulatory standards is unknown. But their elimination from such standards would diminish the time to commercialization.

Whether genetic engineering methods are part of the process of developing and producing a new product or are actually involved in the product itself may be a key distinction. New products for which genetic engineering simply facilitates the development and production, but the products themselves are naturally occurring, will receive much less regulatory scrutiny than products for which genetic engineering has resulted in a new organism. For example, animal hormones such as BST will not involve safety testing that is any more complicated than for agricultural chemicals, such as pesticides and animal drugs. In addition, the "food supply" issue for these biotechnology products should not be any different than for other hormones currently in the food supply. Regulatory review of these products may focus on socio-economic impacts on small farmers and rural communities.

Social and economic issues related to new technologies should be used to anticipate the impacts of these new products and advise strategies to ease the adjustment process. Such considerations should not be used to restrict product development. The inevitability of technical change, its concomitant impacts on social and economic relationships, and the importance of new technologies for international competitiveness will necessitate a regulatory policy that emphasizes science-based safety criteria in decisions to restrict new product development.

NOTE

1. The views expressed are the authors' and do not necessarily represent policies or views of the U.S. Department of Agriculture.

REFERENCES

Alexander, M. 1985. Ecological consequences: reducing the uncertainties. *Issues in Science and Technology* 1:57-68.

American Medical Association. 1935. Dangers of lead arsenate as spray in orchards. *Journal of the American Medical Association* 67:68-79.

Animal and Plant Health Inspection Service. 1986. AQI Work Accomplishment Data Report, FY 1986. APHIS/PPQ, Hyattsville, Maryland.

Annexstad, J. 1986. Bovine somatotropin controversy and impact. *Dairy Herd Management* 45:22-26.

Borcherding, J. R. 1987. Growth hormone work gets a health checkup. *Successful Farming* 26:30.

Bromfield, K. R. 1980. Soybean rust: some considerations relevant to threat analysis. *Protection Ecology* 2:251-57.

Cavers, D. F. 1939. The food, drug, and cosmetic act of 1938: its legislative history and its substantive provisions. *Law and Contemporary Problems* 21:34-67.

Conservation Foundation. 1985. Biotechnology: how tight must our control be? *Conservation Foundation Letter*. May-June:5.

Conway, R. 1982. An economic perspective on the California Mediterranean Fruit Fly Infestation. Staff Report No. AGES820414. Washington D.C.:USDA-ERS.

Council for Agricultural Science and Technology. 1977. Pesticides for minor uses: problems and alternatives. Report No. 69, CAST, Ames, Iowa.

_____. 1981. Impact of government regulation on the development of chemical pesticides for agriculture and forestry. Report No. 87, CAST, Ames, Iowa.

Doyle, J. 1985. Altered harvest: agriculture, genetics, and the fate of the world's food supply. New York: Viking Press.

Dunlap, T. R. 1978. The triumph of chemical pesticides in insect control, 1980-1920. Environmental Review. 26:23-45.

_____. 1981. DDT: scientists, citizens and public policy. Princeton University Press.

Goldhammer, A. 1986. Perspectives on EPA regulation of biotechnology products. Speech presented at the Washington International Conference on Biotechnology, Alexandria, Virginia, April.

Hardy, W. F. and Glass, D. J. 1985. Our investment: what is at stake? Issues in Science and Technology 1:69-82.

Jaffe, G. A. 1987. Inadequacies in the federal regulation of biotechnology. Harvard Environmental Law Review 11:491-549.

Jones, D. D. 1985. Commercialization of gene transfer in food organisms: a science-based regulatory model. Food Drug Cosmetic Law Journal 40:477-493.

_____. 1986. Legal and regulatory aspects of genetically engineered animals. In Genetic engineering of animals, eds. J. W. Evans and A. Hollaender. Washington, D. C.: Plenum Publishing Corporation.

Kenney, M. 1985. Biotechnology conference targets regulatory questions. Genetic Engineering News, June 6:5.

Levin, S. 1986. The risk assessment/confidentiality balance. Paper presented at the Washington International Biotechnology Conference, Alexandria, Virginia, April.

Levin, S. and Harwell, M. A. 1985. Environmental risks and genetically engineered organisms. In Biotechnology: implications for public policy, ed. Sandra Panem. Washington, DC: The Brookings Institution, 1985.

Marcus, A. 1980. Environmental Protection Agency. In The politics of regulation, ed. James W. Wilson. New York: Basic Books, Inc.

McGaritt, T. O. 1985. Regulatory biotechnology. Issues in Science and Technology 1:40-56.

McGregor, R. C. 1973. The emigrant pests. A re-Staff report. Washington D.C.: USDA-APHIS.

Pesticide and Toxic Chemical News. 1987. Biotech subcommittee's work complicated by political reality. March 4:1-2.

Pfadt, R. E. 1971. Fundamentals of applied entomology. 2nd ed. New York: The Macmillan Company.

Regal, P. J. 1987. Models of genetically engineered organisms and their ecological impact. Recombinant DNA Technical Bulletin, Vol. 10, No. 3, September. Publication No. 87-899.Washington D.C.: USDHHS-NIH.

Rifkin, J. 1984. Algeny: a new word--a new world. New York: Penguin Books, Inc.

Rohwer, G. G. 1985. Regulatory plant pest management. In Handbook of pest management in agriculture, ed. D. Pimentel, pp. 253-96. Boca Raton: CRC Press, Inc.

Schneider, K. 1986. Biotechnology regulations are signed by Reagan. The New York Times. June 19:1.

_____. 1986. Eased rules on gene engineering weighed. The New York Times. May 22:1.

Schwartz, P. H. and Klassen, W. 1981. Estimate of losses caused by insects and mites. In Handbook of pest management in agriculture, ed. D. Pimentel, pp. 15-77. Boca Raton: CRC Press, Inc.

Sill, W. H., Jr. 1982. Plant protection. Ames, Iowa: The Iowa State University Press.

Simberloff, D. 1981. Community effects of introduced species. In Biotic crisis in ecological and evolutionary time, ed. H. Nitecki, pp. 53-81. Academic Press.

Thompson, P.B. 1987. Agricultural biotechnology and the rhetoric of risk: some conceptual issues. Paper FAP87-01. Washington D.C.: Resources for the Future.

Vandenbergh, M. P. 1986. The rutabaga that ate Pittsburgh: federal regulation of free release biotechnology. Virginia Law Review 72:1529-68.

Wechsler, A. E., J. E. Harrison and J. Neumeyer (consultant). 1975. Evaluation of the possible impacts of pesticides legislation on research and development activities of pesticide manufacturers. Washington D.C.: EPA-Office of Pesticide Programs, February.

4. The Debate over the Deliberate Release of Genetically Engineered Organisms: A Study of State Environmental Policy Making

In the last ten years the U.S. industry has invested in excess of 5 billion dollars in the hope that scientific developments in biology will yield new commercial opportunities (Kenney 1986). The industries most involved will be pharmaceuticals, agriculture, energy, forestry, waste processing, and even mining. In most of these applications, with the exception of pharmaceuticals it will be necessary to introduce new genetically modified organisms into the environment [1]. This commercial imperative has led to debate regarding the types and stringency of regulation necessary to ensure public safety.

Most parties to the debate agree that some sort of "regulation" is required, but in the last five years no precise regulatory scheme has been introduced. The correct protocols for measuring problems, and even the exact jurisdictional boundaries of various government agencies, have not been developed. Many company scientists hold the conviction that there are no safety hazards at all in biotechnology. However, in 1985 a regulatory framework began to evolve. Many industry groups now recommend the abolition of governmental control, except where it assists in securing liability insurance or prevents local jurisdictions from regulating.

The theories of the state and the state's role in capitalist society have become an increasingly important topic in sociology (see, for example, Skocpol 1985). However, among sociologists the bulk of the research has concentrated upon labor and social welfare policy. Regulation policy has largely been left to political science and economics. This chapter discusses the theories of these researchers with reference

73

to the current debate about deliberate release. The
first section will briefly characterize the major
theories of regulation and the factors each theory
regards as most salient. Then a short history is
provided of the biotechnology industry and its precon-
troversy regulatory regime. This is followed by an
examination of the debate regarding deliberate release
and the types of institutions being developed to regulate introduction of genetically engineered organisms
into the environment. To properly understand the issues involved it is necessary to examine the various
opinions expounded by scientists regarding safety. The
evidence is then reviewed to evaluate which theoretical
model of regulation provides the most persuasive explanation. Conclusions are drawn regarding the likely
regulatory regime to be employed.

THEORIES OF REGULATION

Regulation is an important form of participation
by the state in the production process. For example,
as early as the 1850s, the English state began to regulate the use of child labor in factories. In regard to
the regulation of genetically modified organisms, the
issue is the state's involvement in ensuring that the
production process does not produce what the economists
term "externalities." Put in a more prosaic fashion,
regulation is being considered to ensure that released
organisms cause no unanticipated deleterious side effects. In regulating, the state is claiming, in effect, the right to forbid the production and use of
certain products, if the costs to the entire society
are deemed too high.

The traditional argument in capitalist states for
justifying regulation of corporate activity is that the
state is acting in the public interest, i.e., an interest greater than the individual interest. This position appears intuitive when an agency such as the
Environmental Protection Agency (EPA) orders removal of
toxic chemicals from the environment or invokes its
legal authority to ban the use of certain chemicals.
And, in fact, the EPA is charged by its charter with
protecting the "public health and safety" from toxic
substances. To accomplish this goal, Congress provided
the EPA with police powers to force compliance. The

regulation of the biotechnology industry would appear to be a classic case of the government acting in the public interest.

Public interest theory has in the last twenty years increasingly been challenged. Other perspectives argue that the regulatory agencies often are actually captured by the interests of the very groups they were appointed to regulate. There are two distinct variants to the "capture" hypothesis that we will examine. The first is best exemplified by Gabriel Kolko (1963; 1965) who argues that economic regulation of industries such as the railroads was actually favored and even encouraged by the railroad industry.

Kolko argues that the Interstate Commerce Commission and the other regulatory agencies created in the Progressive Era were fashioned by industry to alleviate or exclude harmful competition and, in certain instances, to co-opt popular demands. This stands in sharp contrast to the traditional beliefs that regulation was instituted to serve the public interest. Rather, regulation is invoked by industries seeking to protect themselves from ruinous competition or demands by popular classes. The state, in effect, enters the marketplace to protect industry from harming itself. Yet, the justification for intervention is that the state is operating in the "public interest."

Another variant of a "capture" thesis was propounded by Bernstein (1955) and was further documented by public interest groups in the 1970s. This argument was that regulatory agencies, commissions, etc. are initially zealous but then become more procedural until they eventually become a tool of industry. The history of regulatory bodies is metaphorically understood as a life cycle. As the regulatory body matures, procedures become increasingly formalized, and influence is wielded by the full-time industrial lobbyist. This occurs because the smooth operation of the commission usually requires a modicum of cooperation from the regulated industry.

A final model that has been presented interprets the forces that led to a regulatory regime in a more open political conflict model. This model of the creation of regulatory agencies is summed up by McCraw (1975, p. 182): "Seldom in American history did the goals of private groups form a perfect identity with those of the rest of society, but seldom a perfect antithesis, either. Within the zones of overlap, private groups plausibly claimed service to society, and capture coexisted in fleeting calm with 'public interest.'"

Wood (1985) demonstrates that the 1906 Food and Drug Act enacted to control adulteration and misla-beling of products was not simply a victory for popular protest, but also was actively supported by important capitalists who were in the process of trying to con-solidate the food and drug industries and control com-petition. As Wood (1985, p.432) indicates:

> The act's most obvious strength lay in its appeal to both consumers and businesses as pro-tection against the inefficiencies and abuses of free-market competition. But the law also ad-dressed the interests of businesses suffering from nothing more than competition itself. In both symbolic and substantive ways, the law does seem to have been a victory for all -- except, of course for adulterators, misbranders, and those whose newer products threatened the market posi-tions of more established firms....The law's passage effectively removed the issue of food and drug adulteration from the public agenda for a number of years. Free from public scrutiny, manufacturers could cultivate access to adminis-trative decision-making channels in the federal government and attempt to secure ad hoc, unpub-licized exemptions from the law.

This model of group and even class compromise is far more sophisticated than the "capture" model, while not denying that capture may occur. It does not deny the obvious fact that elites may have an extraordinarily strong influence on the regulatory body creation pro-cess. Conversely it recognizes the importance of other groups or classes in the process and recognizes that their interests can be served or appear to be served.

The "state" that emerges in this model is a state that is somewhat inactive until the pressure of events becomes so overwhelming as to compel action. The ac-tions taken, however, are not random. Rather, the actions are built upon previous proposals, or in a more general way, on an agenda set by social actors. State bureaucrats, contrary to Skocpol's (1985) arguments regarding their relative autonomy, are far less impor-tant than the social pressures and coalitions to which

they respond. They appear to act autonomously until
one investigates the social actors behind the legisla-
tors or state bureaucrats.

The debate concerning regulation of deliberate
release of genetically-altered organisms can be pic-
tured as taking place on a social battlefield with
various persons and groups trying to maximize their
personal and their institutional positions. This "war
of position" however is inherently unstable because the
technology and other social forces are undermining old
positions. Further, it is intrinsic to any new technology that there will be unknown benefits and costs to
actions thereby making certainty impossible.

GENETIC ENGINEERING

The current debate concerning the deliberate release of genetically engineered organisms into the
environment is a continuation of an earlier debate
regarding the safety of genetic engineering and some of
the protagonists remain the same.[2] The first recombinant DNA (rDNA) debate was initiated by molecular
biologists themselves. The success by Cohen and Boyer
in genetically engineering an organism opened an entirely new technological vista. Simultaneously, a
number of molecular biologists became concerned over
possible risks of the new technology and prompted a
moratorium on experiments until their safety could be
established.

In the wake of the self-imposed
voluntary moratorium an international group of
molecular biologists gathered at Asilomar, California
to discuss the safety issues. They issued a call to
the National Institutes of Health, the major founder of
rDNA research, to draw up health and safety guidelines.
In the interim period before the guidelines were
created, the public was also becoming aware of both the
commercial   potentials   and   possible   risks.
Extraordinary press coverage raised public fears of
possible "monsters" and other unlikely possibilities
from scientific laboratories. What scientists had
planned as an "among friends" discussion of risks,
rapidly became a public furor with calls for stringent
and expensive regulation. Congressional .pa
hearings on the possible risks to public safety of rDNA

research were held every year 1975 through 1978. However, the regulations dealt only with recipients of federal monies. [3]

As the principal funding agent for molecular biology, the National Institutes of Health (NIH) created the Recombinant DNA Advisory Committee (RAC) in response to the scientists. RAC was composed of scientists, government bureaucrats, and token public representatives. RAC undertook to evaluate rDNA experiments and advise the director of NIH regarding their safety. RAC had no legal powers and was given no enforcement powers except in the case of federal research grant recipients. By its makeup RAC was a classic case of capture being largely composed of the very scientists it was supposed to oversee. The stated role of RAC guidelines was to ensure that any potentially dangerous genetically engineered organisms did not escape into the environment. The largely unstated role was to allay public fears (see, for example, Krimsky 1982).

Simultaneously, knowledge concerning the real risks of rDNA was increasing and ever larger numbers of experiments were exempted from seeking RAC approval. By 1982, RAC nearly voted to make its guidelines entirely voluntary. However, some of the more farsighted members remarked that RAC served the important purpose of reassuring the public that its safety was being protected. Thus, the reason for RAC's retention was to provide the public with the perception that it was being protected and thereby preclude public pressure for state and local governments to implement further regulatory efforts.

Initially RAC activities only concerned scientists, but from 1978 onwards, industry took greater interest in rDNA regulatory activities. Industry, though not required to abide by RAC rules, decided that it was in its interest to conform, as the guidelines were not onerous and caused no significant changes in procedure. However, by 1980 it had become clear that RAC, though very useful as a "regulator" for experimentation, would clearly not be suitable for pharmaceuticals. In pharmaceuticals the Food and Drug Administration (FDA) clearly had jurisdiction, and the FDA intended to regulate biotechnology products in exactly the same way as other pharmaceuticals.

DELIBERATE RELEASE -- INDUSTRY DEMANDS REGULATION

The deliberate release of genetically modified organisms into the environment for any purpose was a very different problem and required different regulators. The initial rDNA debate was settled by assuring the public that genetically engineered organisms would not escape and survive in the environment. Providing this assurance was acceptable to scientists and industry because the products then under consideration were pharmaceuticals and basic chemicals, both of which could be produced under containment in fermentation vessels. However, the other major genetic engineering applications such as agriculture, forestry, and waste processing usually require using the living genetically modified organism as a product to be introduced into the environment.

To understand the debate and halting movement by the federal government to regulate deliberate release, it is first necessary to understand the extent of knowledge that was available regarding the risks associated with introducing new living organisms into the environment. The role of the state in aiding industry in its drive to commercialize products also provides insights.

Risks Concerning Deliberate Release

The evaluation of the risks of releasing genetically engineered organisms is difficult because ecological science is not deterministic. Ecological systems are so complex as to defy causal modeling. Accurate evaluation also is plagued by the endless interactions between different components of the environment. Ecologists cannot predict with certainty any outcomes of introductions because effective analysis and research capability is currently unavailable. Thus, much of the debate has been conducted on the basis of case studies of other exotic introductions in the hopes that these will assist in providing safety parameters for the proposed releases (for a chronology, see Appendix A). The lack of knowledge and concomitant uncertainty among scientists is profound.

This scientific debate features molecular biologists (many of whom are employed by commercial entities) and ecologists (many of whom receive Table

4.1

Probability of a Deleterious Effect from a Genetically
Engineered Organism is the Product of Six Factors

---

1. Will the organism be released? P_1

2. Will the released organism survive? P_2

3. Will the surviving organism multiply? P_3

4. Will the organism be transported to a site
    where it will have an effect? P_4

5. Can the genetic information coded in
    organism be transferred to other species? P_5

6. Will the organism be harmful? P_6

$$P = P_1 \times P_2 \times P_3 \times P_4 \times P_5 \times P_6$$

Source: Adapted from Alexander (1985, p.63).

federal funding for their research and wish to see it
increased). Martin Alexander (1985), a Cornell University microbial ecologist, has proposed a scheme to
evaluate the risks of a deliberate release (Table 4.1).
The molecular biologists general argument is that there
is no (or very small) probability of the creation of
harmful organisms. Also, any harmful organisms created
would be at a competitive disadvantage and thus disappear.

The most obvious concern is that a genetically
engineered organism might occupy and become established
in an ecological niche. Deleterious exotic microorganisms introduced into the new environment include the
chestnut blight and the Dutch elm disease, both of
which were introduced into North America from overseas
and quickly attacked and destroyed susceptible native
American species (Sharples 1982). Such devastations
are very difficult to either predict or control and are
what ecologists term high-cost low probability events.
The counterargument from those favoring relatively
unregulated release is that genetic engineering directed at producing useful products would not result in

constructing pathogens inadvertently.

There is less concern regarding the possibility of a new organism occupying an ecological niche while not immediately disturbing the surrounding ecosystem. However, conclusions about apparent technological safety are relative. In the chemical field an important example of this is DDT. Few suspected that DDT could concentrate in birds and operate to inhibit the pro duction of egg shells, which endangered a number of avian species (Alexander 1985, p.116). With living organisms it is difficult to predict far-removed indirect effects.

The use of the ice minus bacteria was proposed in separate experiments by Stephen Lindow and Advanced Genetic Sciences. In the current debate, an objection has been raised that the ice nucleation bacteria may be involved in the formation of raindrops. In a science fiction scenario, it is hypothesized that the release of massive numbers of ice-minus bacteria could change rainfall patterns (Odum 1985, p.1338). This unlikely scenario is illustrative of the unanticipated problems that might result from deliberate release.

Another area of dispute concerns the ability to unwittingly produce a new or more virulent weed through genetic engineering. Winston Brill (1985 p. 382), president of research at Agracetus, has argued:

> From our growing understanding of the genetic and biochemical basis of competition by weeds, it is obvious that many genes must interact appropriately for the plant to display the undesirable properties of a weed (efficient seed dispersal, long seed viability, rapid growth in an environment not normally favorable to other plants. . . . It should be even more difficult to derive such a weed through acquisition of characterized foreign genes.

Waclaw Szybalski (1985, p.115), a cancer specialist at the University of Wisconsin and long-time "DNA chauvinist," was even more confident in a letter to Science: "Brill refers to risk as 'very small' or 'extremely unlikely,' instead of saying they are nonexistent from a practical, societal point of view" [emphasis in the original]. Dr. Eugene Fox (1985, p. 238), director of the ARCO Plant Cell Research Institute, sarcastically mocked ecologists:

> Regulations must be written to prevent muta-
> tion, unnatural exchanges of DNA in nature, and
> indeed evolution itself (enforcement admittedly
> will be a problem). Does anyone else detect the
> musty aroma of the Luddites?

These objections characterize the positions of molecular biologists and corporate personnel as they attempt to win the regulatory policy debate.

The position of ecologists is not that there is a
definite risk but that there could be adverse impacts,
and society should take steps towards developing a
capability to evaluate those risks. Further, as
numerous ecologists have argued, there is little if any
research available to substantiate any position in the
safety debate (Alexander 1985; Sharples 1982; Colwell
1985).

In a letter to Science, Colwell et al. (1985,
p. 11) answered Brill's article first of all by criti-
cizing his description of the "properties of a weed."
Citing as an example the world's most serious agricultural weed, purple nutsedge, which "thrives under
the same conditions as 52 crops in 92 countries and
propagates almost exclusively by vegetative means, and
not by seeds" (rather a distinct contrast to Brill's
emphasis on seeds). As importantly, they note that
many commercial crops outcross with weeds, thus genes
moved to crop plants could possibly end up in a weed.
The outcome could be to rapidly transfer exotic genes
to weeds . In an exhaustive search of the literature
on the characteristics of weeds, Vidaver (1985, p. 170)
concludes:

> The implications for environmental release
> of genetically engineered crop plants are that
> the potential for escape of desirable traits into
> weedy species will be substantial when billions
> of plants planted annually are the result of
> genetic engineering. The more such crops are
> planted, the more likely it becomes.

Traits such as pesticide resistance are being
engineered into crop plants. However, this resistance
may provide the weed with a competitive advantage and
even can cause crop losses. The point ecologists make
is that problems could occur. Thus, releases should
proceed cautiously and increased research concerning

possible predictive methods of assessing the dangers of a new genetically engineered product for release should be developed.

Regulation is Explored

Ralph Hardy and David Glass (1985) among others have complained that genetic engineering is the first technology to be regulated before it has caused a problem. Their point appears to be that the public should wait for a problem before being concerned. [4] Simultaneously, Hardy and others are interested in having the government "regulate" biotechnology to quiet public doubts.

Industry representatives have repeatedly said that they do not want new laws passed that might mandate rigid regulations that could not be loosened if no accidents occur. The gist of industry's argument is that the current laws concerning the agricultural pesticide industry can also handle biotechnology. However, they have simultaneously complained that the development of a consistent regulatory framework from a patchwork of laws prepared for other purposes was too slow.

In 1980 the only institution concerned with rDNA oversight was the RAC which was composed largely of molecular biologists and fermentation experts, i.e. those familiar with production utilizing contained organisms. The question of wanting to release organisms into the environment had not arisen. However, in March 1980 Ronald Davis of Stanford submitted a proposal to NIH requesting permission to test genetically engineered maize plants in the field (i.e., the first deliberate release). Although this proposal was approved at the June RAC meeting, the director of NIH announced in the Federal Register the deferral of any action on the proposal until the USDA approved the experiment.

The necessity of providing some sort of regulatory regime for the environmental release of genetically manipulated organisms was first broached in a 1981 Office of Technology Assessment (OTA) report. It offered Congress the following options for addressing the risks associated with genetic engineering:

1. Maintain the status quo by letting NIH and the regulatory agencies set federal policy.
2. Require that the Federal Interagency Advisory Committee on Recombinant DNA Research prepare a comprehensive report on its members' collective authority to regulate rDNA and their regulatory intentions.
3. Require federal monitoring of all rDNA activity for a limited number of years.
4. Make the NIH Guidelines applicable to all rDNA work done in the United States.
5. Require an environmental impact statement and agency approval before any genetically engineered organism is intentionally released into the environment.
6. Pass legislation regulating all types and phases of genetic engineering, from research through commercial production.
7. Require NIH to rescind the Guidelines.
8. Consider the need for regulating work with all hazardous microorganisms and viruses, whether or not they are genetically engineered (OTA 1981, pp. 230-34).

With these alternatives the OTA outlined the important issues and set the standard of discussion on risks. But the outcome among these alternatives has been guided by industry; neither Congress nor the executive branch have been galvanized to propose any regulations. This is probably because the U.S. political system responds to political pressure, and thus far, little widespread political pressure by citizens has occurred.

The OTA report, however, stimulated thought in the private sector concerning the type of regulatory regime that would be most conducive to their interests and plans. Various federal agencies such as the EPA, the USDA, and the FDA began to ponder their possible role in regulation. With its cursory reviews and lack of legal power, NIH was the most attractive regulator in this early phase. NIH also was reluctant to adopt an enforcement role (for which it was obviously ill-equipped), in addition to its traditional role as a research funder.

In 1981 very few modified organisms were ready to be field tested and the biotechnology industry had only recently weathered the debates of the 1970s. Few agencies felt any compelling necessity to develop special regulatory-like programs for rDNA. Concomitantly, there was little perceived necessity to attempt to clarify boundaries between the various regulatory agencies. Further, the Reagan administration was philosophically opposed to regulation, ensuring that the voluntary NIH Guidelines and RAC would set the standards. However, as Karny (1981) pointed out, to not secure the appropriate approvals through RAC would likely open the experimenter to charges of negligence under tort law and, of course, the possibility of a lawsuit. A further concern was the fact that insurers were ill-disposed to provide coverage for experiments that had no upper limit on the risk (Chakrabarty 1983).

In 1982 two further experiments were referred to NIH for action: one from John Sanford of Cornell University to test genetically altered plants and one from Stephen Lindow of the University of California, Berkeley, to field-test genetically modified bacteria for the purpose of modifying plant frost resistance (ice minus or INA-). This research was supported by the venture capital-financed biotechnology startup company, Advanced Genetic Sciences, Inc., which had exclusive commercial rights to the bacteria. However, the approved plans to test the genetically engineered plants were discontinued by the investigators because of technical problems and only the ice minus investigators continued with plans to field-test (U.S. Congress 1984, p. 177). The October 1982 RAC meeting voted 7-5-2 to allow the ice minus release, but approval was withheld until more information was supplied to the RAC.

In October 1982 an official report (written the previous year by Frances Sharples, an Oak Ridge National Laboratory and 1981 AAAS EPA fellow) was approved that evaluated the risks of deliberate release. It recommended that the EPA take a role in regulating deliberate releases under both TSCA (Toxic Substances Control Act) and FIFRA (Federal Insecticide, Fungicide and Rodenticide Act), though it was recognized that these laws would have to be stretched to encompass deliberate release (Karny 1981; U.S. Congress 1984).

The EPA was not the only agency that could plausibly claim jurisdiction over at least certain aspects of deliberate release. For example, the Plant Quarantine Act, the Federal Noxious Weed Act, the Federal Plant Rest Act and Virus, Serum, and Toxin Act provide the

USDA with the power to control any agents that might be harmful to agriculture. Finally, the USDA is empowered through Executive Order 11987 to regulate any (deliberate or otherwise) releases of exotic organisms into the environment (U.S. Congress 1984, p. 36-39). With these powers it would seem that the USDA would be in an especially powerful position to regulate biotechnology. However, the USDA was unwilling to prepare to take a role in regulation.

Even as the EPA began to consider requiring industry to secure permits for testing, Ralph Hardy, director of Life Science Research for Du Pont Chemical, testified to Congress that there was no need for regulation:

> Such a controlled approach to beneficial applications of the results of new recombinant technology can be covered through the existing oversight mechanisms contained in the NIH Guidelines. We support the NIH Recombinant DNA Advisory Committee's (RAC) position with review by the local Institutional Biosafety Committee and a Subcommittee of the RAC (Hardy 1983, p. 93).

Ralph Hardy was expressing the view of most industry spokesmen, which was that there was no reason to develop any mandatory regulations for industry. There was, in fact, no regulation. Industry did not even have to inform RAC of its activities, if it chose not to. Further, RAC from its inception obliged industry by permitting all hearings on industry applications to be entirely confidential (Fox and Norman 1983, p. 1355) [5]. And, of course, numerous RAC members were intimately involved with industrial funding. The RAC method of regulation was more akin to licensing than regulation. A U.S. Congressional (1984, p. 27) staff report summarized the possible problems very succinctly:

> It is private industry, not research institutions, that has the greatest desire to move ahead with the release of new organisms on a large scale in order to achieve the broadest commercial applicability of their new genetically-engineered products. Given the highly competitive nature of the biotechnology industry, it is not unlikely that some companies may decide not to request RAC

approval out of a concern that the approval process could cause costly delays in testing and
marketing their products.

RAC's voluntary compliance program was subject to easy
circumvention.

THE COURT AS A FIELD FOR GUERILLA WARFARE

With the RAC approval, Lindow and the University
of California were preparing to undertake the first
field test of genetically engineered organisms when the
social activist, Jeremy Rifkin, and a number of other
parties filed for an injunction in the Washington Federal District Court on September 14, 1983. This legal
move petitioned the court to require that NIH prepare
an environmental impact assessment regarding the release of the ice-minus bacteria (Foundation on Economic
Trends 1983). Rifkin's suit brought quick reactions as
the EPA accelerated its plans to develop regulations
for deliberate release; approval under TOSCA by the EPA
would eliminate the necessity of an environmental impact report. However, due to Rifkin's suit the field
test was cancelled and rescheduled for spring 1984.
Bowing to congressional pressure and increasing industry concerns, the Office of Management and Budget as-
sembled in March a working group of fifteen federal
agencies to sort out the various agencies' regulatory
responsibilities (Sun 1984b, p. 855). This was fortui-
tous. Judge John Sirica stunned NIH on May 16, 1984 by
a ruling that suggested that Rifkin's complaint had a
good possibility of success, and he issued an injunction against NIH from approving any further field tests
by academic scientists (Norman 1984).
Rifkin's court success prompted a more active
search by industry for a regulator and also called into
question the role of RAC in approving field tests. For
example, Monsanto announced that it would go directly
to the EPA for field test approval of its newly engineered bacteria; this led to the acceptance of the EPA
as the regulator of genetically engineered organisms
(Sun 1984a, p. 697). This acceptance by industry in
August 1984 of the EPA's role was in sharp contrast to
less than a year earlier. At that time, Donald Clay,
the acting administrator of the EPA Office of

Pesticides and Toxic Substances, claimed that companies had threatened to sue if the EPA regulated deliberate release under TSCA (Sun 1983, p. 823).

The ideology of regulation as activity in the public interest is illustrated in the case of deliberate release. The EPA could serve a useful role and in 1984 Donald Clay was quoted as saying, "I'm looking to regulate with a light hand," as he and other EPA officials agreed with industry that it was unnecessary to offer any new legislation (Sun 1984a, p. 6). Clay clearly indicated the political nature of the EPA's regulatory activity when he says that industry was very sensitive to regulation but it was necessary to keep public confidence in government protection (Hanson 1984, p. 38). Dr. Will Carpenter, Monsanto Agricultural Products Co. general manager for technology, states his viewpoint:

> Regulation is needed, first because the public and environment must be protected. Second, and equally important, the public must perceive that they and the environment are being protected. They must have confidence that work is done by responsible scientists with the approval of responsible regulators (Journal of Commerce 1984, p. 228).

The emphasis was not on public safety, but rather the appearance of safety. If Clay had any inclination to be tough, it was quashed by OMB which, at the suggestion of Monsanto, intervened in the EPA's regulation development (Sun 1984b, p. 472). This apparent lack of commitment to rigorous standards and large-scale research on potential problems created questions regarding the seriousness of federal regulatory efforts. Environmental activist Jack Doyle charged that the regulations being implemented were more in the nature of "pseudoregulation" (Doyle 1985).[6]

Edward Jefferson, chairman of the board of Du Pont, devoted his speech at the opening of an $85 million Du Pont life science laboratory to a call for federal regulation:

> Recombinant DNA research and development of a biotechnology industry are proceeding in a climate of regulatory and legal uncertainty... It is unac-

ceptable to leave biotechnology in a regulatory limbo (<u>Chemical and Engineering News</u> 1984, p. 6).

Jefferson said those "in industry must be prepared to take an early and clear position in favor of working closely with government on these matters," and he outlined five recommendations for government action:

1. Develop a matrix that will establish which products fall in which agencies' jurisdictions.
2. Develop rational, scientifically justified guidelines.
3. Establish an interagency committee as a sounding board for interested parties.
4. Appoint a special counselor on biotechnology for the president.
5. Use NIH RAC as a successful template for regulation (<u>Chemical and Engineering News</u> 1984, p. 6).

However, Jefferson's greatest concern was any stringent regulations that might retard the rush to the market.

Ralph Hardy, in his new position as president of Biotechnica International Inc., recommended that the federal government use existing laws, create no new requirements, have one agency for each area, use "proper" scientific review, use the risk record from traditional biotechnology, do more risk research, and bear the costs of regulation (Hardy and Glass 1985, p. 81). Even while Congressman Dingell was castigating the EPA and the USDA for not adequately developing research tools for understanding the possible impacts of genetically engineered organisms, the EPA was rushing forward to "regulate" (Dingell 1984). On December 31, 1984 the Office of Science and Technology Policy (OSTP) published a notice in the <u>Federal Register</u> (1984) declaring the manner and legal justifications by which the FDA, the EPA, and the USDA would regulate biotechnology.

Sorting out regulatory responsibility was made difficult by the curious lack of USDA interest in specifying any regulatory mechanism, although it had relatively clear statutory jurisdiction (Colwell 1985). In the December 31, 1984 <u>Federal Register</u> notice, the USDA states its regulatory philosophy, "USDA anticipates that agriculture and forestry products developed by modern biotechnology will not differ fundamentally from conventional products" (OSTP 1984, p. 50898). Hanson (1984, p. 37) also remarked that the USDA was taking its role far less seriously than the other agencies.

A General Accounting Office (1985) report found that the USDA seemed unclear on what to do and which internal branch had which regulatory responsibilities. Karim Ahmed (1985) charged this was due to the fact that the USDA had always accepted its role as being dedicated to yield increase rather than regulation of technology. Nonetheless, its lackadaisical attitude towards preparing a concrete plan of regulation had contributed to the delays in the development of guidelines as well as regulatory uncertainty.

As Jeremy Rifkin effectively tied up NIH through lawsuits by securing a judgment that environmental assessments were required. The EPA route became more appealing to industry.[7] When filing under TSCA, it was only necessary to file a description of the proposed release. The burden was on the EPA to establish the risks involved with a proposed experiment; the release could take place if the EPA did not respond in 90 days. This is a particularly onerous rule for an already understaffed agency in a field in which the EPA was desperately scrambling to secure trained personnel. However, this particular situation was desirable from industry's viewpoint by providing a public assurance function and possible tort law protection. Any problems that occurred due to deliberate release would legally be placed on the government. Industry could argue that it had fully complied with the law and that any problems were due to inadequate regulation.

The National Academy of Sciences thought that the topic of deliberate release was sufficiently vital to propose a large in-depth study assessing the dangers that might occur with deliberate release. However, there was little interest on the part of the executive branch in funding research concerning the fate of recombinant organisms. For example, the National Academy of Sciences proposed a two year $600,000 study on the risks of deliberate release, but the executive branch's spokesperson on biotechnology, Bernadine Bulkley (1984), testified at a congressional hearing that the administration thought the study was too expensive. Similarly, industry has not funded research available to the public on possible health and safety impacts of their products or contributed to the cleanup of any areas adversely impacted by released bacteria. The position of industry has been communicated very well to politicians, who have also held off on creating any regulatory framework.

On November 11, 1985 the EPA had approved the release of INA- and Rifkin once again responded with a lawsuit. A number of ecologists said they supported the EPA's position, but many questions have remained unanswered (Colwell 1985; Alexander 1985). By late 1986, increasing numbers of products for field-testing were filling the pipeline, but a regulatory schema had still not been fully developed. Further, the necessary scientific protocols had not yet been developed. The expectation can only be that an increasing number of proposals for deliberate release will be encountered, and an overloaded EPA interested in deregulation will approve nearly all proposals. In fact, John Moore (1984, p. 15) testified in 1984:

> While I believe we have an excellent start in addressing the risks of environmental release of genetically engineered organisms, I also believe it is clearly an area where additional research is essential. Development of standard protocols and guidelines for assessing risks would be beneficial in ensuring consistent risk assessment.

Despite these laudable goals, few new research projects have been funded to develop protocols for the approval process or to develop detection methods for rDNA organisms that may be causing problems.

CONCLUSIONS

Biotechnology is an important new technology and federal bureaucrats are developing a regulatory regime that is both desired and demanded by industry. So far the development and deployment of biotechnology is proceeding with little public input. Nevertheless, the success of Jeremy Rifkin in conducting what might be characterized as a legal "guerrilla" campaign against genetic engineering has created feelings of paranoia among corporate executives. The weakness in Rifkin's campaign has been that until recently he has had little success mobilizing citizen participation.

The demands of industry -- that the regulations be based on laws developed for petrochemicals -- have been quickly adopted, even though there is little justification for treating living organisms as chemicals. Even these small efforts to regulate occurred only after Rifkin's suit succeeded in requiring that NIH conduct environmental assessments on all federally-funded releases. The effectiveness and timing of the new EPA regulatory regime are suspect since it is obvious that the EPA still has little in the way of expertise regarding rDNA. The TSCA regulatory regime is lax and places the burden of proof on an EPA that does not want to regulate. It seems obvious that in the current regulatory climate little concrete protection will be offered to the public.

The current relatively stringent review of proposals will probably be relaxed as public attention becomes focused on other issues. In effect, in this debate we see the process by which a "captive" regulator has been created at the EPA. In the case of deliberate release, the "public interest" paradigm offers little explanatory power. In fact, most public representatives and legal counsel warned industry that the public required the "perception" of regulation and that real regulation was not necessary. Industry and government were urged to cooperate to supply the public with exactly that illusion.

The "capture" theories of regulation do provide a somewhat more robust explanation of the reality of rDNA regulation. The probusiness Reagan administration has been forced to provide regulation due to public pressure, however, they wanted to create captive regulators. The most obvious manifestation of this pressure has been Jeremy Rifkin's lawsuits, which have used the legal edifice of environmental laws built up by the struggles of the 1960s and 1970s to stymie corporate will for nearly two years. Rifkin's rearguard actions are now becoming more desperate. It is ironic that now local communities where microorganism releases are planned, are beginning to resist this activity in their jurisdictions. It may be the nature of the current U.S. system that the federal government increasingly is seen as the preserve of business, while struggles against new technologies take place at the local level.

The destruction of the so-called onerous regulations of the federal government may unleash a complex wave of contradictory local regulations creating a

problematic situation for large companies that want
national uniformity for their planning process. Thus,
the "free enterprise" state (that aims not only at
securing the capture of state bureaucracies for industry, but also their outright surrender) may result in
destroying the "uniform playing field" that is essential for industry. To secure the uniform playing field
it is probable that the EPA will make regulations more
stringent in an attempt to preempt state and local
regulations. The focus will be on creating an illusion
of regulation while trying to ensure that local jurisdictions are unable to make their own laws.

This chapter examines the accepted view that regulation is conducted in the "public interest." State
managers are seen to be constantly deferring to representatives of the private sector and attempting to
secure their support for initiatives. The policy-
making process is deliberately secretive, but nevertheless, evidence suggests that political economic
analysis using state theory can yield predictions regarding the outcome of social struggles. In fact, the
development of clear biotechnology regulations will
probably not occur until public pressure forces governmental response.

Appendix A. Chronology of Events in the Debate over
Release of Genetically Engineered Organisms into the
Environment

_____

_____

3-80	Request to RAC by Ronald Davis of Stanford for NIH permission to test genetically transformed maize.
6-5-80	RAC votes to permit Davis experiment.
7-29-80	Director of NIH announced in Federal Register that he will defer action on Davis experiment.
4-81	Office of Technology Assessment report Impacts of Applied Genetics mentions possible problems with deliberate release.
8-7-81	NIH announces in Federal Register that it will approve Davis field test.
6-9-82	Request by John Sanford of Cornell University to field-test altered plants.

9-22-82   First NIH _Federal Register_ notice of application for ice minus bacteria field test from Stephen Lindow of University of California, Berkeley.

10-82   Sharples report to the EPA concerning possible problems with deliberate release completed.

10-25-82   RAC meeting votes 7-5-2 to allow ice minus release (and approval withheld).

2-23-83   The USDA RAC meets and approves Sanford proposal.

3-3-83   The EPA begins to look at regulation.

3-4-83   _Federal Register_ gives notice of deliberate release.

3-8-83   Lindow brings back revised proposal.

4-11-83   RAC approves revised Lindow proposal 19-0.

4-15-83   Sanford experiment gets final approval for test.

6-1-83   Lindow experiment receiving permission to field-test appears in _Federal Register_.

6-22-83   House hearings on deliberate release.

9-14-83   Rifkin files first suit in Federal District Court against NIH for approving ice minus without EIS.

3-11-84   The EPA claims jurisdiction over deliberate release.

3-12-84   The OMB calls for Cabinet-level council to review rDNA regulatory needs.

4-12-84   Rifkin files for preliminary injunction against University of California.

5-9-84   First working group meeting of 15 federal executive agencies potentially involved in regulation.

5-16-84   Preliminary injunction against University of California/NIH granted by Judge Sirica.

6-8-84   House Agriculture Committee has hearings on deliberate release.

8-5-84   Monsanto announces it will go to the EPA and not RAC.

12-84   Monsanto informs the EPA it wants to test an rDNA organism.

12-31-84   Proposal for coordinated framework _Federal Register_ notice involving the EPA, the FDA, the USDA, and NIH published.

1-21-85   NIH Environmental Assessment completed and it finds no problems.

1-25-85	Science publishes long article from Winston Brill saying there are no safety concerns regarding deliberate release.
2-27-85	Appeal of District Court injunction prompts Appeals Court to lift injunction on other releases but approves injunction on Lindow release.
3-15-85	AGS withdraws petition to NIH and sends to the EPA.
4-15-85	NIH Federal Register notice soliciting comments on Lindow experiment environmental impact assessment.
9-27-85	Odums letter in Science responding to Brill and arguing there might be ecological cost.
11-14-85	The EPA approves field test of ice minus (AGS).
11-14-85	Rifkin files complaint asking for declaratory and injunctive relief against the EPA.
11-15-85	Monsanto makes BT submission.
12-20-85	The EPA returns Monsanto submission with comments for revision.
1-2-86	Monsanto returns to the EPA again.
1-16-86	Rifkin seeks preliminary injunction against the EPA for approving ice minus.

NOTES

1.   Some in industry have argued that because there is natural transfer of genes there are no new genetic recombinations (Brill 1985).   It should be noted that this observation is not included in patent applications that refer to "novel organisms" that are the result of genetic engineering.

2.   Genetic engineering refers to the ability to add and delete genes in the DNA molecule and then to insert the DNA into a cell and have it expressed.   The first successful recombinant DNA or genetic engineering experiment was done by scientists from Stanford University and the University of California, San Francisco in 1973.   Since then scientists have been successful transferring genes from species to species and then having those genes expressed.

3.   For further information on the first recombinant DNA debate see Krimsky (1982); Lear (1978); Watson and Tooze (1981).

4.   This strategy of waiting until a problem occurs before taking certain insurance and monitoring measures typified U.S. regulatory activities during the last 40 years in the petrochemical and pharmaceutical industry.   Now the public faces tremendous cleanup bills.   It is interesting to note that many of the companies now urging regulatory "restraint" with regards to biotechnology are among the worst polluters in petrochemicals.

5.   This secrecy was the focus of an unsuccessful lawsuit by Jeremy Rifkin.

6.   Jack Doyle being of the opinion that regulation has some objective criteria, rather than being largely political.

7.   The environmental assessment of the Lindow experiment was remarkably short and indicated no serious effort to monitor the released organisms (National Institutes of Health 1985).

REFERENCES

Ahmed, K. 1985.   Government oversight of biotechnology. Environmental and Energy Study Conference, U.S. Congress.

Alexander, M. 1985. Ecological consequences: reducing the uncertainties.   Issues in Science and Technology Spring:57-67.

_____. 1985. Planned releases of genetically-altered organisms: the status of government   research and regulation.   Hearing before the Subcommittee on Investigations and Oversight of the Committee on Science and Technology, U.S. House of Representatives. 4 December. Washington, DC: U.S. Government Printing Office.

Bernstein, M. 1955.   Regulating business by independent commission.   Princeton, NJ: Princeton.

Brill, W. 1985.   Safety concerns and genetic engineering in agriculture.   Science 227 25 January:381-4.

Bulkley, B. 1984.   Biotechnology regulation.   Hearing before the Subcommittee on Oversight and Investigations of the Committee on Energy and Commerce House of Representatives. 11 December. Washington, DC: U.S. Government Printing Office.

Chakrabarty, A. 1983. Environmental implications of genetic engineering. Hearings before the Subcommittee on Investigations and Oversight and the Subcommittee on Science, Research and Technology of the Committee on Science and Technology, U.S. House of Representatives. 22 June. Washington, DC: U.S. Government Printing Office.

Chemical Week. 1985. 'Ice minus' bacteria gains EPA approval. Chemical Week 27 November:98.

Chemical and Engineering News. 1984. Biotech regulation: Du Pont chief urges clearer policy. Chemical and Engineering News 24 September:6-7.

Colwell, R. 1985. Planned releases of genetically-altered organisms: the status of government research and regulation. Hearing before the Subcommittee on Investigations and Oversight of the Committee on Science and Technology House of Representatives. 4 December. Washington, DC: U.S. Government Printing Office.

Colwell, R.; Norse, A.; Pimentel, D.; Sharples, F.; and Simberloff, D. 1985. Genetic engineering in agriculture. Science 229 12 July:111-2.

Dingell, J. 1984. Biotechnology Regulation. Hearing before the Subcommittee on Oversight and Investigations of the Committee on Energy and Commerce House of Representatives. 11 December. Washington, DC: U.S. Government Printing Office.

Doyle, J. 1985. Government oversight of biotechnology. Washington, DC: Environmental and Energy Study Conference, U.S. Congress.

Foundation on Economic Trends et al. 1983. Foundation on economic trends et al. vs Margaret Heckler et al. Complaint filed in the United States District Court for the District of Columbia. Civil Action No. 83-2714.

Fox, J. 1985. Genetic 'engineering'. Science 230 18 October:237-8.

Fox, J., and Norman, C. 1983. Agricultural genetics goes to court. Science 30 September:1355.

Government Accounting Office. 1985. Biotechnology: the U.S. Department of Agriculture's biotechnology research effort. GAO #RCED-86-39BR.

Hanson, D. 1984. Government readies rules for biotechnology control. Chemical and Engineering News 13 August:34-38.

Hardy, R. 1983. _Environmental implications of genetic engineering_. Hearing before the Subcommittee on Investigations and Oversight and the Subcommittee on Science, Research and Technology of the Committee on Science and Technology, U.S. House of Representatives. 22 June. Washington, DC: U.S. Government Printing Office.

Hardy, R., and Glass, D. 1985. Our investment: what is at stake? _Issues in Science and Technology_ Spring:69-82.

Journal of Commerce. 1984. Regulation held a must for biotechnology. _Journal of Commerce_ 11 June:228.

Karny, G. 1981. Regulation of genetic engineering. _University of Toledo Law Review_.

Kenney, M. 1986. _Biotechnology: the university-industrial complex_. New Haven: Yale.

Kolko, G. 1963. _The triumph of conservatism: a reinterpretation of American history, 1900-1916_. Princeton, NJ: Princeton.

_____. 1965. _Railroads and regulation, 1877-1916_. Princeton, NJ: Princeton.

Krimsky, S. 1982. _Genetic Aaachemy: the social history of the recombinant DNA controversy_. Cambridge, Mass.: MIT.

Lear, J. 1978. _Recombinant DNA: the untold story_. New York: Crown.

Moore, J. 1984. _Biotechnology regulation_. Hearing before the Subcommittee on Oversight and Investigations of the Committee on Energy and Commerce House of Representatives. 11 December. Washington, DC: U.S. Government Printing Office.

McCraw, T. 1975. Regulation in America: a review article." _Business History Review_ 492:159-83.

National Institutes of Health (NIH). 1985. Environmental assessment and finding no significant impact. NIH, Public Health Service, 21 January.

Norman, C. 1984. Judge halts gene-splicing experiment. _Science_ 224 1 June:962-3.

Odum, E. 1985. Biotechnology and the biosphere. _Science_ 229 27 September:1338.

Office of Science and Technology Policy (OSTP). 1984. Proposal for a coordinated framework for regulation of biotechnology, notice. _Federal Register_ 31 December:50856-907.

Office of Technology Assessment (OTA). 1981. _Impacts of applied genetics_. Washington, DC: GPO.

Sharples, F. 1982. Spread of organisms with novel genotypes: thoughts from an ecological perspective. Oak Ridge National Laboratory, Environmental Sciences Division, Publication No. 2040. October.

Skocpol, Jheda. 1985. Bringing the state back in strategies of analysis in current research. In Bringing the state back, eds. P. Evans, D. Rueschemeyer, T. Skocpol, pp. 1-43. Cambridge: Cambridge University Press.

Sun, M. 1983. EPA revs up to regulate biotechnology. Science 223 18 November:823-4.

_____. 1984. Biotechnology's regulatory tangle. Science 225 17 August:697-8.

_____. 1984. White House enters fray on DNA regulation. Science 224 25 May:855.

Szybalski, W. 1985. Genetic engineering in agriculture. Science 229 12 July:112-3.

U.S. Congress. 1984. The environmental implications of genetic engineering. Staff Report prepared by the Subcommittee on Investigations and Oversight transmitted to the Committee on Science and Technology, U.S. House of Representatives. February. Washington, DC: U.S. Government Printing Office.

Vidaver, A. 1985. Plant-associated agricultural applications of genetically engineered microorganisms: projections and constraints. Recombinant DNA Technical Bulletin 83:97-102.

Watson, J., and Tooze, J. 1981. The DNA story. San Francisco: Freeman and Company.

Wood, D. 1985. The strategic use of public policy: business support for the 1906 Food and Drug Act. Business History Review 59 Autumn:402-32.

Part II

Shifts in Farming and Agriculture

5. Theories of Technical Change in Agriculture with Implications for Biotechnologies

Technology is one of the most potent forces affecting agriculture, both from historical and contemporary perspectives. In the developed world, technology has caused the supply of food to expand faster than demand, precipitating severe adjustment problems in the agricultural sector. In developing countries, where chronic food shortages exist and increases in traditional inputs have proven ineffectual in expanding food supply, technology is often viewed as an "engine of growth" capable of alleviating the more serious problems of starvation and malnutrition.

The United States has had the most productive farming industry in the world. Agricultural output has doubled since 1940 while total input has remained relatively constant. The rate of U.S. agricultural productivity growth, however, began to slow in the 1960s. Lu (1983) reports that from 1939 to 1960, total agricultural factor productivity increased 2 percent annually, but from 1960 to 1970 this rate of increase slowed to .9 percent annually and since has rebounded only slightly to 1.2 percent annually.

The technologies that made this productivity growth possible were primarily energy-related. Tractors, irrigation, pest control chemicals, and fertilizers are all fossil-fuel based. Tweeten (1986) points out that machinery improvements have been responsible for many of the productivity advances. Thus, much of the previous progress has relied on cheap and abundant energy sources. In addition, these technologies may entail societal costs such as pollution, groundwater contamination and depletion, soil erosion, and salini-

103

zation from irrigation. The societal risks associated with biotechnology are still subject to speculation (Molnar and Kinnucan 1985).

A number of global needs and conditions appear to portend the advent of a new frontier of productivity advance in agriculture. Computer and microelectronic devices certainly will enhance farm machinery performance, but much of the future improvements may stem from the built-in properties of plants and animals themselves. Less reliant on fossil-fuels, the new products emanating from genetic engineering may substitute organism-performance for external energy inputs (Molnar, Kinnucan, and Hatch 1986). It is not clear, however, how this rapidly evolving set of technologies will be implemented by food producers (see e.g., Molnar, Kinnucan, and Hatch 1986).

This chapter reviews selected theories of technical change in light of the new biotechnologies. The points of influence and divergence among these perspectives will be treated as a basis for understanding a fundamental force shaping the future of American agriculture. Biotechnologies represent an array of supply-increasing productivity improvements. The variety and rapidity of discovery and farm-level delivery of these products will have marked effects on the future of American agriculture.

THEORIES OF TECHNICAL CHANGE

In examining the theories of technical change and what they might imply for the new agricultural biotechnologies, it may be useful to have in mind a definition of technical change. Definitions abound, but economists would generally agree with the one given by Peterson and Hayami (1977) as "the phenomena of input quality improvements or an increase in knowledge leading to an increase in output per unit of input...." In other words, technical change makes it possible to produce more output with the same quantity of resources previously employed. The result is lower per unit costs of production and therein lies the incentive to adopt new technology, especially in market-oriented economies where prices are free to adjust in response to changes in supply and demand.

Three theories of technical change in agriculture that are applicable to biotechnology are the treadmill theory, the induced innovation theory, and the diffusion theory. Each of these perspectives views technology from a distinct vantage point: the treadmill theory seeks to explain the structural consequences of new technology on agriculture; the induced innovation theory is concerned with the forces that govern the development of new technologies; and the diffusion theory focuses on the technology adoption and dissemination process. A brief summary of each of these theories is presented below.

Treadmill Theory

The treadmill perspective seeks to explain adjustments in agricultural industry as new technology is assimilated by farm firms (Cochrane 1979). The theory argues that farmers are on a treadmill with respect to new technologies because profits initially realized by early adoption of the technology are eventually lost as output expands and prices fall. Late adopters, who have rushed to adopt the technology as a cost-cutting measure in response to declining prices, exacerbate the problem by adding to supply, which in turn accelerates the downward trend in prices. Non-adopters or late adopters are forced into bankruptcy as the technology-induced decline in prices results in a sustained period of financial losses.

When the adjustment is complete, a new price equilibrium is established in which surviving producers earn no profit other than a normal return to land, labor, capital, and managerial ability. The rewards for the decision to innovate are short-term adoption rents for early adopters and avoidance of losses or insolvency.

To regain profitability producers must stand ready to aggressively adopt the next technical innovation that becomes available. Producers who adopt early enough will realize savings on costs of production. For a time prices will remain at pre-innovation levels, permitting early adopters to profit. Eventually, however, industry supply expands and sets off a new round of price decreases that continues until profits are eliminated. Thus the cycle repeats itself and the treadmill continues to turn.

The above depiction describes adjustments that occur in agriculture as technology is introduced into an environment free of government intervention. In the long run, the big winner from such a process is the consumer: he/she receives the same amount of food or more but at a lower price. The losers are the laggards in technology adoption: they find it increasingly difficult to maintain solvency in the face of declining product prices and eventually face bankruptcy and loss of the farm.

The more aggressive innovative farmers who had the prescience to adopt the new technology early also stand to gain in two respects: (1) they realize the "adoption rents" associated with early adoption and (2) they can improve their wealth position by acquiring the assets, particularly farmland, of the late or non-adopting farmers. Thus, although the cost of adjustments to new technology in agriculture can loom very high for the unfortunate laggards, it is quite clear that society as a whole has much to gain when technical change takes place in an unfettered free market economy. The gains will, of course, be un-equally distributed among producers themselves.

However, much of agriculture is not a free market -- most major crops in the United States have a price floor maintained by the government. In this case the scenario depicted by the treadmill theory of continu-ally declining real product prices cannot occur. This fact begs the question "Does the treadmill theory ade-quately characterize the technology-induced adjustment process that takes place in agriculture when prices are not permitted to fall because of government fiat?" Cochrane answers this question by introducing the con-cept of a "land market" treadmill, which he distin-guishes from the "product market" treadmill described above. From this perspective, the early adopting farmer enjoys an especially long period of sustained excess profits because prices do not decline or decline only moderately in response to expanding output. These profits provide a powerful incentive for the early adopting farmer to expand output by acquiring addi-tional land. Land price inflation sets in as early adopting farmers bid for the scarce resource.

Land prices continue to rise until the economic profits originally produced by the new technology are completely eliminated. Because land suitable for agricultural production is essentially fixed in supply, additions to one's land base can be made only by acquiring land from other farmers and these are typically the weak and inefficient farmers who are latecomers to the treadmill. Thus, the acquisition of land by aggressive early adopters becomes a type of "cannibalism" (to use Cochrane's colorful terminology) in which financially stronger farmers consume the assets of their weaker counterparts. Cannibalism then becomes an integral feature of the agricultural adjustment process to new technology when government price supports exist in commodity markets.

The ultimate outcome of the land market treadmill is fewer farms, an increase in the scale of surviving farms, and higher prices for farmland. Under this scenario, the lower per unit production costs originally made possible by the new technology change are completely offset by the higher costs of production caused by rising land prices. Thus consumers fail to gain, laggards lose as usual, and the only beneficiaries are surviving (early adopting) farmers. Thus governmental intervention in markets sometimes reduces the long-term benefits offered to society by new technology. In other ways, support programs reduce downside price risk and encourage investments in new technology.

Induced Innovation Theory

Whereas the treadmill theory seeks to explain adjustments in agriculture that occur in response to the availability of new technologies, induced innovation theory attempts to shed light on why technology develops in the direction that it does. What causes a technology such as hybrid corn, the mechanical tomato picker, or the cotton gin to be developed?

First proposed by an economist named J. R. Hicks in 1932, the theory of induced innovation identifies changes or differences in the level of relative prices as the key force influencing the direction of innovative activity and hence the direction of technical progress. According to Hicks, "The changed relative prices will stimulate the search for new methods of production which will use more of the now cheaper factor and less of the expensive one." Thus, for example, if labor becomes more expensive relative to capital, scientists and engineers will be encouraged to seek ways to economize on the more scarce input by developing new labor-saving machinery.

Some disagreement exists as to whether changes in the relative costs of production inputs really influence the direction of technical progress. For example, labor costs may be high relative to capital costs, but if labor-saving knowledge is more costly to generate than capital-saving knowledge, there is no reason to expect technology to develop in the direction of labor-saving innovation (Salter 1960). Furthermore, if per unit labor costs are high relative to per-unit capital costs, individual firms will seek labor saving technology because this is the most effective way to gain cost reduction (Kennedy 1974). In any case, the level of input costs, and not their relationship to one another, is all that is essential for a theory of induced innovation (Peterson and Hayami 1977):

Induced innovation theory, originally developed to explain technological progress in the private sector, has since been extended to public sector investment in new agricultural technologies (Hayami and Ruttan 1971). Because many of the new technologies developed in agriculture are the product of publicly-sponsored research, this extension seeks to establish a link between relative factor prices in agriculture and the research efforts. The nexus is described by Hayami and Ruttan (Peterson and Hayami 1977):

> Farmers are induced, by shifts in relative prices, to search for technical alternatives which save the increasingly scarce factors of production. They press the public research institutions to develop the new technology, and also demand that agricultural supply firms supply modern technical inputs which substitute for the more scarce factors. Perceptive scientists and science administrators respond by making available new technical possibilities and new inputs that enable

farmers to profitably substitute the increasingly abundant factors for the increasingly scarce factors, thereby guiding the demand of farmers for unit cost reduction, in a socially optimum direction.

A pivotal link in the public sector inducement mechanism, as described above is the response of research scientists and administrators. The extent to which these individuals respond to local clientele groups accelerates the adoption process.

Diffusion Theory

Whereas the treadmill theory made mention of early adopters and laggards without attempting to differentiate between them, diffusion theory focuses on how and why people (mainly farmers) respond to new technologies. The object is to increase understanding of the processes whereby new technology is assimilated into the production activities of firms. The approach has been principally one of determining the factors that explain differential rates of adoption and regional diffusion of new technology.

Early studies done by rural sociologists and geographers tend to focus on communication and sociocultural resistance as factors determining the pattern of diffusion over time and space (Beal and Bohlen 1957). Familiarity with a technique or the ability of farmers to observe the effects of new technology and convey these impressions to others are cited as critical factors in explaining the rate of adoption (Havens and Rogers 1961). Later work by economists points to such factors as the perceived profitability of the innovation (Griliches 1960, 1962), the asset position of the firm (Mansfield 1970), and the skill distribution of potential adopters (Huffman 1974).

Ilbery (1985) cites eight important generalizations about the links between decision-making and innovations:

1. The creation of new knowledge and practices largely occurs outside farms. Very few farmers are inventors of new practices.

2. A large number of channels of communication
    exist to inform and advise farmers on new
    innovations, ranging from salesmen to gov-
    ernment advisory services. However, farmers'
    receptivity to them varies widely.

3. Individual farmers are in a weak marketing
    position and tend not to conceal ideas from
    one another. Consequently, interpersonal con-
    tact between farmers can be an important chan-
    nel of information. But farmers in a region
    are far from uniform and much diversity exists
    in terms of farm type, scale of operation, and
    farmer characteristics.

4. New information and innovations take time to
    be adopted by the farming community (and some
    never are).

5. The need to become informed and consider inno-
    vative ideas arises from farmers' motivations
    and problems. Once again these vary consider-
    ably, reflecting differing perceptions among
    farmers of the qualities and characteristics
    of new ideas.

6. The possibility of rejection or non-adoption
    must be part of any model that aims to be
    descriptive of a farmer's decision-making.

7. The alternatives open to a farmer are much
    wider than the simple acceptance or rejection
    of the idea. Courses of action available are
    related to the amount and quality of knowledge
    possessed by farmers, which varies enormously.

8. Any new information is surrounded by uncer-
    tainty and farmers will look for a satisfac-
    tory solution among possible alternatives. An
    important component of the decision-making
    process in this context is the variation in
    socio-personal characteristics of the farmers.

Farm size is often considered an important determinant of technology adoption based on the tendency
of early adopters to come from the ranks of larger farm
operators. An economic rationale for this apparent empirical regularity is offered by Feder and Slade (1984)
who argue that larger farmers face a lower cost of
adoption because they can spread the fixed cost of
learning about the new technology over a greater level

of output, thus achieving greater relative reductions in per unit cost. However, Just and Zilberman (1988) note that farm size may be serving as a surrogate for other factors affecting technology adoption, such as the ability to bear risks, financing capability, and risk preferences.

In explaining diffusion of technology across geographically-distinct regions, the notion of adaptive research has been shown to play a critical role (Hayami and Ruttan 1971). Agricultural technologies often are location specific and do not respond well in new unintended environments without adaptive research. Failure to recognize the location-specific character of agricultural technology is sometimes referred to as a major factor in explaining difficulties encountered when attempting international technology transfer (Peterson and Hayami 1977). Griliches (1957), in his efforts to explain regional differences in the rate of adoption of hybrid corn in the United States, incorporates a mechanism of local adoption that takes the location-specific character of the technology into account.

Ilbery (1985) identifies a number of empirical regularities that have been observed in studies of the diffusion of innovations:

1. The hierarchical effect, or tendency for individuals in large places to adopt earlier than people located farther down the hierarchy.
2. The neighborhood effect, or tendency for diffusion to be strongly influenced by the friction of distance. Therefore, within the hinterland of a single urban center, diffusion is expected to proceed in a wave-like fashion outwards, first affecting farmers in the rural-urban fringe rather than in purely rural areas. A similar pattern is expected in diffusion among a rural population. Various studies, at different geographical scales, have been concerned with the neighborhood effect, including Griliches's (1957) study of diffusion of hybrid corn in the United States and Hagerstrand's (1967) study of bovine tuberculosis control in Sweden.
3. The logistic effect, or the tendency of the

cumulative level of adoption over time to approximate an S-shaped curve. The actual shape of this curve depends on the nature of the innovation and the degree of resistance expressed by potential adopters.

4. A fourth possible category is the <u>random effect</u>. Spatial regularities in diffusion patterns do not always exist or may be overshadowed by selective awareness and subsequent adoption associated with media participation, marketing efforts, or links to widespread social networks.

Another aspect of diffusion theory considers infrastructure and market forces affecting the introduction and adoption of new technologies (Brown 1983). Infrastructure refers to the network of agencies and firms that makes an innovation available and supports its use. Firms selling new innovations develop marketing strategies that can effectively isolate segments of the market from the technology because sales potential in these areas are considered limited. (In the case of hybrid corn, for example, seed companies sent representatives to visit personally farmers in Iowa, Illinois, and other corn belt states whereas farmers in the South and other areas where corn production is more dispersed had to learn about the product on their own initiative. Griliches (1957) termed this the "availability problem.")

Establishment of diffusion agencies (e.g., marketing organizations in the case of private sector innovation and extension services or outreach programs for public sector innovation) may be a precondition for adoption. Thus an understanding of the supply side forces affecting technology diffusion is crucial to a holistic view of the diffusion process. The implication is that a portion, if not much, of the observed differences in the spatial patterns and temporal rates of diffusion can be explained by institutional (as opposed to individual) behavior.

The amount, type, and quality of information available to farmers for decision-making is central to understanding how the geography of agriculture is created and how patterns of production might be reinforced or altered (Ilbery 1985). One study of inter-regional contrasts in farm publications found magazines to be the most important sources of information for farmers. Generalized farm magazines tended to be published in a small number of metropolitan centers whereas specialized magazines tended to be published near the areas within the region or growing area they endeavor to serve (Ilbery 1985). Such specialized sources also are an important source of specific information on new production technology and may contribute to the previously mentioned random diffusion pattern, particularly when the innovation requires few supporting inputs or services.

IMPLICATIONS FOR AGRICULTURAL BIOTECHNOLOGIES

Biotechnology innovations are arriving in a bewildering variety of products throughout the many industries in the agricultural sector. The specificity of these products is perhaps the most distinguishing attribute of the genre termed biotechnology. Specific products must be understood in the context of their performance characterisitics, their relative profitability, and the regulatory environment controlling their release and application.

Product Performance

The relative advantage of biotechnology innovations will have a great deal to do with their spread in an industry. Easier, safer, more convenient, less costly, and more effective substitutes for existing products will hold sway in the marketplace. Relative advantage is a subjective evaluation reflecting the constellation of product traits vis-a-vis the individual producer's own preferences and situation. Clearly the greater the required deviation from existing production practice, the greater the resistance to be expected from traditional farmers. Direct substitutes for existing products that have visibly superior performance characteristics will be the most readily adopted innovations.

One current biotechnology product with clear performance advantages is a growth hormone called bovine somatatropin (BST) for use in dairy cattle. Researchers report 10 to 20 percent gains in milk production with only secondary modifications to feed regimes. In a recent study, Southeast dairy farmers were surveyed concerning their potential adoption of BST (Hatch, Kinnucan, and Molnar 1985). The decision to adopt was disaggregated into four steps: awareness, feasibility, adoption, and intensity of adoption. Herd size was found to be significant in explaining awareness and adoption.

Many observers suggest that BST is management intensive and will not necessarily be suitable for large dairy operations. More productive, innovative farmers are expected to be the early adopters of BST. In the survey, productivity and use of artificial insemination were significant in explaining awareness and potential adoption. Successful use of artificial insemination, a previously adopted new technology, may influence potential adopters' active information gathering behavior. Adoption has been characterized as a learning process in which present period experience is added to previous period opinion. Adoption occurs as a result of a dynamic information gathering process; thus, the decision to acquire information is an endogenous element (Feder and O'Mara 1982). If a dairy farmer has been pleased with the results of previous new technologies, he or she is more likely to adopt.

Induced innovation theory may suggest implications relative to regional impacts to the extent that location specific adoptive research might be needed to tailor BST or other new biotechnologies to regional resource endowments. Recent production research indicates that BST is less effective when animals are under stress from heat or humidity (Kalter 1985). Milk deficit areas in the South and West may adopt early to circumvent transportation costs incurred by importing from the northern tier of dairy exporting states. This shift in favor of the South will occur if preliminary research results overstate the stress factor in reducing effectiveness or if biotech firms are induced to modify BST to suit regional needs. Size economies that indicate that one firm will be capable of fulfilling industry demand for BST suggest that regional tailoring will not be profitable in the case of BST.

Profitability

Profitability clearly drives adoption behavior by U.S. farmers. The pricing decision for the producer of the new technology is "How much of the margin between the cost of the new product and the benefits to farmers will be captured?" Innovations are typically higher-priced at the time of introduction when supply is limited, particularly for clearly advantageous products. The entry of competitive substitutes and accumulated production economies often leads to lower prices for later adopters, however. Furthermore, not all firms have the same standing in the marketplace relative to the benefits of an innovation.

Treadmill theory suggests that early adopters capture the economic rents associated with a new technology. Early adoption has been associated with farm size in many studies of technology diffusion. Will biotechnologies, and BST in particular, exhibit this size bias? Several considerations underlie the impact of size.

First, the global nature of the size dimension dictates a need to disaggregate its component variables. Size may be capturing several effects, including fixed adoption costs, risk preferences, human capital, credit constraints, labor requirements, and tenure arrangements. To the extent that a new technology represents a fixed cost, economic logic dictates a bias in favor of larger farms.

Although many biotechnical innovations may represent strictly variable costs, gathering the information on which the decision to adopt is based represents a fixed cost (Feder and Slade 1984). Consequently, even these technologies can work to the disadvantage of smaller farms. Empirical results, generally obtained in connection with high-yielding varieties in less developed countries, have tended to support the hypothesis that larger farmers adopt earlier. Ruttan asserts that although many innovations initially are adopted by larger farmers, smaller farmers eventually do adopt. If the agricultural treadmill theory is valid, early adopters will capture the economic rents associated with biological innovations and later adopters will be forced to use the improved technology to maintain solvency (Cochrane 1979).

The biotechology products that are currently being marketed or are near commercialization tend to have low fixed costs for the adopter. The most important fixed cost may be information acquisition. While the potential exists for information acquisition to add a fixed adoption cost, its relative importance to determining the scale neutrality of new biotechnologies is unclear.

Regulatory Environment

Fear and uncertainty over biotechnology innovations present hurdles to product development and use. Elaborate but necessary procedures for field testing increase costs and delay introduction. Public fears may be exacerbated by a single unfortunate accident or misstep covered by existing rules (Perrow 1984). Nevertheless, additional constraints may increase the justification necessary for safety certification and marketing. Theories of diffusion pertain primarily to the uptake of an advance once it becomes available. Current societal trends operate to lengthen the delay of entry to the marketplace. Increased costs associated with certification also impede diffusion by raising prices and reducing profitability. Thus, natural diffusion processes are clearly constrained by regulatory steps that delay and delimit the use of new products in the interest of safety and the protection of resources.

CONCLUSION

The course of biotechnology innovations on American farms can be understood through theories of innovation generation, adoption, and diffusion. These perspectives point to the importance of competitive processes in fueling and rewarding the demand for innovations. Market and infrastructure approaches emphasize supply constraints on adoption behavior. New products are not uniformly available to potential adopters, nor are the information sources that lead to awareness and use of new biotechnologies evenly distributed. Farmers in clusters or agglomerations of similar producers have advantages in time and in technical support over isolated or disparate producers of the same commodity. Public extension services and private consultants often are more attuned to the needs of enterprises in specialized production areas.

Adoption behavior is influenced by firm characteristics as well as individual traits. Resource availability and receptiveness to risk are likely to be greater on larger farms.

The broad concept of management ability can be seen as facilitating the adoption of productive, profitable innovations. The component variables of education, training, farm magazine readership, and certain personality traits are more specific explanations for adoption behavior.

Finally, the institutional regulatory environment can be seen as a staging area for innovations before market mechanisms, social networks, and diffusion agencies can make products available and facilitate their use. The future of biotechnology is likely to be most greatly influenced by the political economy of regulation than the relative reluctance or receptivity of individual farmers to innovations. Natural processes associated with profitability and relative advantage will convey biotechnology innovations to farmers when the products become available (albeit sooner for some than others). The real concerns for farmers and consumers pertain to the rate at which new products are allowed to be developed and tested, the speed at which efficacy and safety can be determined, and the level of confidence that all sectors of society have in the whole system.

REFERENCES

Beal, G. M. and J. M. Bohlen. 1957. The diffusion process. Iowa State Agricultural Experiment Station Report 18.

Brown, Lawrence A. 1983. Diffusion of innovations. New York: Nethune.

Cochrane, C. C. 1979. The development of American

<u>agriculture</u>. Minneapolis, Minnesota: University of
Minnesota Press.

Feder, G. and O'Mara, G. T.   1982.   On information
and innovative diffusion: a Bayesian approach. <u>Amer-
ican Journal of Agricultural Economics</u> 63:141-45.

Feder, G. and Slade, R. 1984.   The acquisition of
information and the adoption of new technology.
<u>American Journal Agricultural Economics</u> 66:312-20.

Griliches, Z. 1957.   Hybrid corn: an exploration in
the economics of technological change.   <u>Econmetrics</u>
25:501-22.

_____. 1960.   Consequence <u>versus</u> profita-
bility: a false dichotomy." <u>Rural Sociology</u> 25:354-
6.

_____. 1962.   Profitability versus inter-
action: another false dichotomy. <u>Rural Sociology</u>
27:327-30.

Hagerstrand, T.   1967.   <u>Innovation diffusions as a
spatial process</u>.   Chicago, Illinois: University
of Chicago Press.

Hatch, U., Kinnucan, H., and Molnar, J. J.   1985.
Factors influencing the adoption of new biotechnolo-
gies.   Paper presented to the Annual Meetings of the
American Association of Agricultural Economics, Ames,
Iowa.

Havens, A. E., and Rogers, E. M.   1961.   Adoption of
hybrid corn: profitability and the interaction effect. <u>Rural Sociology</u> 26:409-14.

Hayami, Y., and Ruttan, V. W.   1971. <u>Agricultural development: an international perspective</u>. Baltimore:
Johns Hopkins University Press.

Hicks, J. R. 1932.   <u>The theory of wages</u>. London: Macmillan.

Huffman, W. E. 1974.   Decision making: the role of
education.   <u>American Journal of Agricultural Eco-
nomics</u> 56:85-97.

Ilbery, B. W. 1985.   <u>Agricultural geography: a social and economic analysis</u>. Oxford, England: Oxford
Press.

Just, R., and Zilberman, D. 1988.   Stochastic structure, firm size, and technology adoption.   <u>Oxford
Economic Papers</u> (forthcoming).

Kalter, R. et al. 1985.   <u>Biotechnology and dairy industry: production costs, commercial potential and the
economic impact of the bovine growth hormone.</u> Agri-

cultural Economics Research 85-20, Cornell University, December.

Kennedy, C. 1974. Induced bias in innovation and the theory of distribution. Economic Journal 74:514-17.

Lu, Yao-Chi. 1983. Forecasting emerging technologies in agricultural production. In Emerging technologies in agricultural production, ed. Yao-Chi Lu. Washington, D.C.: USDA-Cooperative State Research Service.

Mansfield, E. 1970 Microeconomics. New York: W. W. Norton Co., Inc.

Molnar, J. J., Kinnucan, H., and Hatch, U. 1986. Anticipating the social impacts of biotechnology in agriculture: a review and synthesis. In Application of biotechnology to agricultural chemistry, ed., H. LeBaron et al. Washington, D.C.: American Chemical Society Monograph Series.

Molnar, J. J., and Kinnucan, H. 1985. Biotechnology and the small farm: implications of an emerging trend. In Strategies for survival of small farms: international implications, ed., T. T. Williams. Tuskegee, Alabama: Human Resources Development Center, Tuskegee Institute, Chapter 1.

Perrow, C. 1984. Normal accidents: living with high risk technologies. New York: Basic Books.

Peterson, Willis, and Hayami, Y. 1977. Technical change in agriculture. In A survey of agricultural economics literature, ed., Lee R. Martin. Volume 1. Minneapolis: University of Minnesota Press.

Rogers, E. 1985. Diffusion of innovations. Third Edition. New York: Free Press.

Ruttan, V. 1977. The green revolution: seven generalizations. International Development Review 19:16-23.

Salter, W. E. G. 1960. Productivity and technical change. Cambridge: At the University Press.

Schmookler, J. 1966. Invention and economic growth. Cambridge: Harvard University Press.

Tweeten, L. 1986. Macroeconomic policy and the structure of agriculture. In Agricultural change - consequences for southern farms and rural communities, ed., J. J. Molnar. Boulder: Westview Press.

Ronald D. Knutson, Robert D. Yonkers,
James W. Richardson

6. The Impact of the Biotechnology and Information Revolutions on the Dairy Industry

The dairy industry is entering an era of exceedingly rapid technological change. As such, it will be the first of the agricultural industries to feel the effects of the coming biotechnological and information revolutions. These simultaneous revolutions have the potential for profoundly affecting the structure of the dairy industry.

Currently, the U.S. dairy industry is composed of two major segments:

1. Traditional family farm dairying is typical of the Northeast, the Ozarks, the Upper Midwest, and most of the Southeast. Dairy producers in these regions generally milk from 30 to 150 cows while farming a sufficient number of acres to supply a majority of the feed inputs used on the farm.
2. Feedlot dairying is typical of the Southwest, the West, and Florida. These operations typically milk from 250 to 5,000 cows with little acreage beyond that required for milking and feeding facilities. Therefore, nearly all feed inputs are purchased.

There is a myth in the dairy industry that the Upper Midwest is the most efficient milk production region in the United States. Since 1973, the USDA cost of production data have shown that the lowest production costs are in California (Figure 6.1), and as a result, the West has experienced a persistently increasing share of national milk production (Betts 1986).

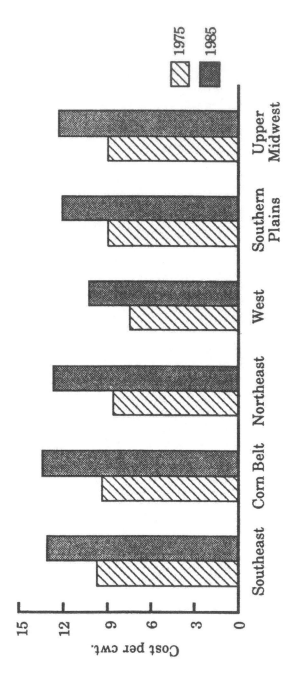

Figure 6.1
Total Costs of Producing Milk in the U.S. by Region, 1975 and 1985

SCOPE OF TECHNOLOGICAL CHANGE
AFFECTING DAIRYING

Technological advances are not new to the dairy industry. Milk output per cow increased 85.4 percent from an average of just over 7,000 pounds in 1960 to over 13,000 pounds in 1985 (Figure 6.2). Most of this increase (about a 2.5 percent annual rate) was due to genetic improvement (largely a result of widespread adoption of artificial insemination technology), improved management ability, and improved quality of forage. Output per cow in 1985 varied widely across the United States ranging from 9,394 pounds (lowest) in Mississippi, 12,900 pounds in Wisconsin, 15,518 pounds in California, to 16,170 pounds (highest) in Washington. These differences reflect a combination of factors including diversity in climatic conditions, management abilities, forage quality, and rates of genetic improvement.

As one looks to the future, four major types of technological change are projected to have an impact on the dairy industry:

1. Recombinant DNA bovine somatatropin (BST) is the synthetically produced version of the major hormone responsible for stimulating milk production. Currently in Food and Drug Administration clearance procedures, BST is expected to be released for commercial use in 1989. A Cornell University study indicates that BST can be expected to increase output per cow by 25 percent or higher in well managed herds (Kalter et al. 1984).

2. Other biotechnologies include innovations in the transfer, freezing, splitting, and sexing of embryos as well as the application of genetic engineering to the development of new disease and pest control systems. The basic characteristic of this technology set is that the impact of milk production is projected to yield a relatively constant 0.5 percent annual rate of increase in output per cow above the past trend (OTA 1986).

3. Nutrition technologies include improved feed delivery systems, use of feed additives, and increases in the quality of forage. These technologies are expected to increase output

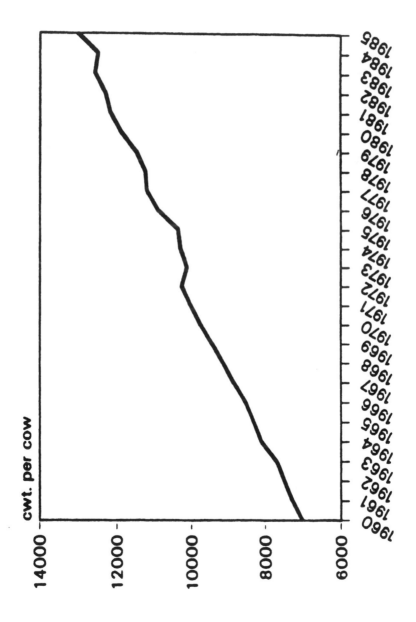

Figure 6.2
Average Annual Milk Production per Cow, U.S. 1960-1985

per cow 0.6 percent annually over the past
trend (OTA 1986).

4. Information technologies are largely com-
    puter controlled milk production and data
    management systems. Such systems will auto-
    matically tabulate output per cow, evaluate
    milk for disease problems, maintain repro-
    duction records, and aid the manager in making
    critical breeding and culling decisions de-
    signed to maximize the herd's genetic value.
    Information technologies have the potential to
    increase milk production per cow 1.2 percent
    annually over the past trend (OTA 1986).

It can be seen from the above
description that all of the technologies except BST are
expected to place relatively constant, upward pressure
on milk production, as opposed to the onetime leap in
output per cow expected after BST is introduced.

ADOPTION PATTERNS

Future technologies will make dairying considerably more complex than in the past. This will create
more technological diversity among dairy producers.
Throughout the past two decades, adoption of dairy
technologies has been relatively uniform across farm
size and region. Most dairy producers use artificial
insemination (AI), milking machines, bulk tanks and --
on an increasing basis -- milking parlors.

In the 1980s, the introduction of embryo transfer
technology and computerized systems began to create a
more diverse industry. Embryo transfer technology
involves the nonsurgical removal of embryos from a
genetically superior female and subsequent implantation
in the uterus of genetically unproven or inferior recipients. The result is an enhanced rate of genetic
progress and, more importantly, the potential for a
larger number of cows with high milk production as well
as bulls with the potential to transfer superior genetic traits. However, recent research indicates that,
utilizing current procedures, embryo transfer technology is clearly not economically feasible for commercial
dairy producers interested only in selling milk -- not
breeding for registered sale (Martin 1985). More ge-

netic gain can be obtained through selection of the sire's genetic path (using the best AI bulls) than through selection of the dam's genetic path, and at a considerably lower cost. The result will be select breeding herds utilizing more advanced embryo transfer, sexing, and splitting technologies while commercial farmers continue to use conventional AI technologies. Commercial use of embryo transfer technologies awaits breakthroughs in embryo splitting and freezing, which will result in a cost per heifer on a level comparable to that of AI alone.

The use of BST will be considerably more widespread, but could also lead to increased industry diversity. Available research suggests that obtaining the full output enhancing impact of BST requires a considerably higher level of management talent. The 10-25 percent increase in milk production is the product of increased feed intake and enhanced energy efficiency by the cow. Feeding practices must compensate for higher nutritional requirements both during lactation and the cow's dry period. Questions thus arise as to how many dairy producers have the management talent to effectively utilize BST.

Dairy producers have expressed reluctance to administer daily injections of BST to their dairy herd, but there appear to be strong economic incentives to adopt BST. Kalter et al. indicate that feed costs increase $4.50 - $5.72 per additional cwt. of milk produced with the BST cost of $0.085 - $0.186 per daily dose per cow, while the government price support level for milk is currently $11.60 per cwt. In addition, by the time BST is introduced, a less frequent injection, implant, or additive could be commercially available.

Information technology likewise has the potential for creating more diversity in the dairy industry. Many dairy producers may be in the position of having neither the resources to own, nor the talent to operate, evolving computer dairy management and control systems. Evolving on-farm information technologies also have some very important implications for traditional off-farm dairy information sources. The National Dairy Herd Improvement Association (DHIA) has been the information repository for individual cow and herd milk production records. Regular on-farm production checks by DHIA of milk production by individual cows are part of the industry record. In addition to these records being used for management decisions, they are the basis for progeny testing of bulls and cows. Prior to the advent of on-farm computer systems, most commercial dairymen found it advantageous to enroll in the DHIA program to receive valuable input for management decisions. On-farm computer systems internalize and customize the information available by recording daily individual cow milk production records, keeping histories of cow health, and aiding in breeding decisions. This development may eliminate much of the need for DHIA to the individual farm. In addition, the genetic records of DHIA will be seriously distorted by individual cow production response to BST, increased industry diversity, and rapid technology advances. For an industry that has thrived on an abundance of records and information, maintaining a consistent data base during technological change will be a major challenge.

STRUCTURAL IMPACTS

From the above description, major structural (if not regional) change in milk production patterns will take place. Instantaneous increases in output per cow provide the potential for large increases in aggregate milk production and for downward pressure on milk prices. The demand for milk is quite inelastic, meaning that there is little potential for increased consumption as price decreases. The government is already purchasing large quantities of manufactured dairy products -- over 10 percent of the milk supply in 1985. As the federal government accumulates even larger stocks of dairy products, pressures will increase to reduce the milk support price.

Given lower milk prices, increasing government stocks, and a lower support price for milk, many dairy producers will be forced out of business. The question is where, why, and what the government might do about this adjustment in the dairy industry?

Where Will the Adjustment Occur?

The 1986 OTA report "Technology, Public Policy and the Changing Structure of American Agriculture," provides considerable insight into the issue of where dairy industry adjustments due to technological change

might occur under a continuation of current policies.
In making projections, this study made some important
assumptions regarding the incidence and rates of adoption of the new technologies. These assumptions are
consistent with the previous discussion concerning
diversity and with the results of available research
on technology adoption in the dairy industry (Kalter et
al. 1984).

It was assumed that the largest dairies in each
region would be the first to adopt the new technologies. These dairies (hereafter referred to as very
large farms) represent the largest 10 percent of the
farms delivering milk to processing plants in each
region. Large farms in the 70th to 90th percentiles
were assumed to adopt late, while moderate-size farms
in the 40th to 70th percentiles were assumed to adopt
still later. Adoption lags varied by technology. The
other biotechnology, information, and nutrition technologies assumed a four-year lag in adoption between
the very large and moderate-size farm, while BST assumed a two year lag.

These lags in adoption rates, combined with the
previously discussed annual rates of change in productivity associated with each technology package, were
analyzed using the farm level dairy simulation model
(DAIRYSIM) developed by Richardson as an expanded version of the Firm Level Income Tax and Policy Simulator
(FLIPSIM V) (Richardson and Nixon 1986). The representative farms used in this analysis were developed by
Buxton for the OTA study. Two farms were included from
the Upper Midwest, three from the Southwest, and three
from Florida.

Some of the major results of this research are
summarized in Table 6.1. In each region, very large,
technically state-of-the-art dairy farms were able to
compete and survive -- although the Upper Midwest 125-
cow farm saw its net worth decline. The farm having
the lowest probability of survival (64 percent) was the
52-cow Upper Midwest dairy, which is typical of moderate-size farms in the traditional dairy regions of
the Upper Midwest, the Ozarks, and to a somewhat lesser
extent, the Northeast. Over the ten year simulation
period, this dairy had an average annual net cash loss
of $11,000 for the farming operation. The implication
from this analysis is that traditional size dairy farms
will bear the greatest pressure to change as a result
of technological advances in dairying.

Table 6.1
Selected Financial Characteristics of Representative
Farms After Ten Year Simulation

Region/ Herd Size	Probability of Survival (a)	Beginning Net Worth	Average Present Value of Ending Net Worth (b)	Average Annual Net Cash Income (c)
	percent	- - Thousands	of dollars	- - -
Upper Midwest				
52 cow	64	417	214	-11
125 cow	100	984	961	22
Southwest				
359 cow	92	745	1,105	41
550 cow	88	1,236	1,572	9
1436 cow	98	2,652	5,931	353
Southeast				
350 cow	86	758	855	13
600 cow	100	1,466	2,250	98
1436 cow	100	3,477	8,671	606

(a) Probability of survival is the probability that a
    farm would maintain an ending equity level greater
    than the minimum required by financial institu-
    tions in the local area.
(b) For farms solvent at end of simulation period.
(c) For all years simulated.
Source: OTA (1986).

Why the Adjustments?

It might be asserted that these results were the
sole product of assumed lags in technology adoption
rates. That is not the case. Subsequent research at
Texas A&M University on the impact of BST alone indicates that while the elimination of adoption lags en-
hances the probability of survival somewhat, it does

not change the basic conclusion that the moderate size
Upper Midwest dairies are the most directly threatened
by technological change and current industry conditions
(Table 6.2) [1].

The primary factor here is the difference in cost
structure among different sizes and locations of dairy
farms. Figure 6.3 clearly indicates that, under projected future technological conditions with the assumed
lags in technology adoption, the moderate size, Upper
Midwest dairy farm has considerably higher costs per
hundredweight. A continuation of these conditions will
lead to a severely depressed dairy economy in traditional dairy regions and intense political conflict
over future dairy policy.

Table 6.2
Probability of Survival After 10 Year Simulation using
BST Technology Assuming the Specified Lags in Rates of
Adoption (a)

Region/Herd Size	No BST	No Lag	Two Year Lag	Four Year Lag
	- - - - - - - - percent - - - - - - - -
Upper Midwest				
52 cow	48	60	58	52
125 cow	100	100	100	100
Southwest				
359 cow	66	74	70	66
1436 cow	76	80	80	80
Southeast				
350 cow	100	100	100	100
1436 cow	100	100	100	100

(a) The probability that a farm would maintain an
    equity level greater than the minimum required by
    financial institutions in the local area.
Source: Yonkers et al. 1986.

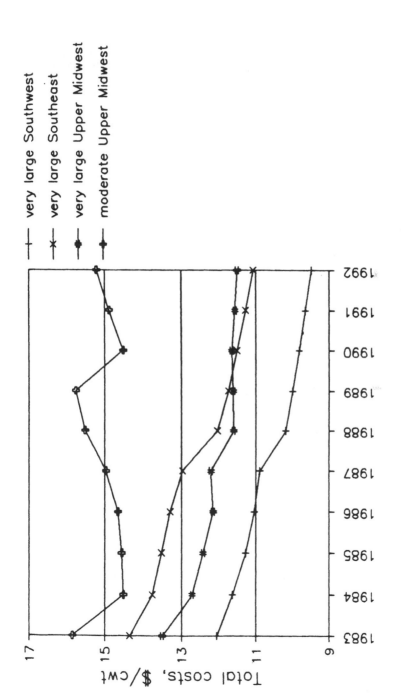

Figure 6.3
Simulated Total Costs of Production in Selected Dairies, 1983-92

GOVERNMENT POLICY OPTIONS

Options are currently being discussed for national dairy policies that previously were considered to be clearly outside the realm of political probability. They are listed here in order, beginning with the most extreme.

Ban BST

At a recent annual meeting of the Minnesota Farmers Union (the state's largest farm organization), a resolution was passed seeking legislative and/or regulatory action to ban BST. Arguments for such action included the current milk surplus situation (and the associated falling milk price leading to increased government expenditure and/or reduced farm numbers) and the inhumane treatment of dairy cows through daily injections and more stressful levels of output per cow. Animal rights activists can be expected to join with such organizations in protesting injections, as they did in protesting the branding of Dairy Termination Program cows.

Government action to prevent technological change is reminiscent of earlier state actions to ban and tax margarine, as well as labor union protests against automation. The BST controversy could become even more heated. Instances in which the U.S. government has banned a technology on economic grounds are rare and counterproductive. In retrospect, butterfat utilization could have been higher today if a butter-margarine blend had been allowed sooner. However, certain farm groups, predominantly in the Upper Midwest have already moved to ban the use of BST in the face of large dairy surpluses. The short-run beneficiaries of such a ban would be the less progressive, status quo dairy producers. The biggest losers would be progressive dairymen, consumers, and the companies holding patent rights on BST.

Production and/or Marketing Quotas

At its 1985 annual meeting, Associated Milk Producers, Inc. (AMPI), the largest dairy cooperative in the United States, voted to seek legislation mandating quotas on the quantity of milk marketed. National mandatory supply controls have never been used for milk in the United States, although considerably less effective base plans are in effect in select markets such as California and South Carolina. Effective, national mandatory controls are in use in Canada under the country's milk marketing board policies.

Mandatory production controls would tend to freeze structural characteristics and regional milk production patterns in their current state. Windfall gains in net worth would accrue to current producers who receive the initial allocation of marketing rights. In Canada, these rights have taken on market values several times the worth of the cows that produce the milk (Fallert and Goodloe 1984). Consumers, of course, wind up paying higher prices as the value of the quota is capitalized into the cost of production. New producers would find entry into the industry more difficult. Traditional dairy regions would benefit relative to expanding production regions.

Periodic Dairy Termination and/or Milk Diversion Programs

In 1984, the government offered to pay dairy producers $10 per cwt. for reducing milk marketings by up to 30 percent. Producers took the government up on the offer to the tune of about 5 percent of 1983 U.S. production. In 1985, milk production exceeded the pre-program level as diversion participants returned their herds to a full production posture.

In 1986, the government paid dairymen up to $22.50 per cwt. of 1985 annual milk marketings to sell their herds for slaughter or export, with the stipulation that the producer not engage in dairying in any form for a period of five years. Dairy producers representing about 12 percent of 1985 U.S. production took the government up on this offer. Most dairy economists and industry experts are predicting that by 1988, production will be at least as high as 1985 levels.

Under both of these programs, the heaviest participation occurred in the highest cost of production region, the Southeast. Interestingly, the Southeast is a milk deficit region and requires expensive interstate trucking of milk. The result is often a processor cost that exceeds local milk production cost. In addition, following the diversion program, Congress increased the price of milk in this region to encourage more production! The consumer in the Southeast pays three times: once for the taxes to pay for the buyout, a second time for higher dairy product costs, and a third time for the government purchases under the price support program that result from the failure to reduce national production.

Price Support Reductions

Price has not been used to control milk supplies since the 1960s. As a result, government purchases of dairy products have gradually built to the point where over $2 billion worth of milk equivalent is purchased annually. Cheese and butter distribution programs have been used to prevent stored purchases from spoiling. These programs do not significantly reduce government stocks, since those who are given cheese then eliminate their purchases of cheese through commercial channels.

Technological change could tremendously accelerate the level of government purchases of dairy products and expenditures. To bring the industry back into a supply-demand balance, price support levels at the time of widespread BST adoption would have to be reduced, maybe by as much as 18 percent ($2.00 per cwt.). Subsequently, the price support level would need to be sufficiently flexible to take into consideration cost changes due to technological advances.

Price would be a harsh adjustment tool. Many dairy farmers retained by past programs would be forced out of business. Government retraining programs would be needed to facilitate their adjustment out of farming. Unless there are bold new efforts to reduce milk production costs in traditional farming regions, the bulk of this adjustment would occur in traditional dairy regions. State governments in support of their Land Grant Universities have a very important role to play in facilitating this adjustment.

SUMMARY

By all measures the dairy industry is on the verge of major structural change. The greatest pressure is going to be on moderate-size dairy farms located in traditional milk production regions such as the Upper Midwest, the Northeast, and the Ozarks. The result will be a political choice between policies designed to prevent change and those designed to facilitate adjustment. Interest groups having the most at stake are the milk producers, the consumers (taxpayers), and the biotechnological and information firms holding property rights to the evolving products and processes. As in any case where the potential exists for government to inhibit or outlaw technological advances based on economic considerations, the scientific community in general has a great deal riding on the outcome.

NOTE

1. This study differs from the OTA study in that only the impacts of BST were modeled. In addition, the representative farms included in this analysis were updated from 1982 to 1985. This period saw a decline in asset values, especially land, and an associated increase in debt ratios. The results of the two studies were not intended to be compared, but rather to address separate research questions.

REFERENCES

Betts, Carolyn. March 1986. Costs of producing milk 1975-1984. Dairy Situation and Outlook. Economic Research Service, United States Department of Agriculture, DS-404. pp. 28-34.

Buxton, Boyd M. June 1985. Economic, policy and technology factors affecting herd size and regional location of U.S. milk production. Paper prepared for the Congressional Office of Technology Assessment. Washington, D.C.

Fallert, Richard F. and Carol A. Goodloe. December 1984. California and Canadian quota plans. Dairy Situation and Outlook. Economic Research Service, United States Department of Agriculture, DS-399. pp.

136

29-36.

Kalter, Robert J. 1984. Production cost, commercial potential and the economic implications of administering bovine growth hormone. Paper read at Cornell Nutrition Conference for Feed Manufacturers. October 1984. Ithaca, New York.

Kalter, Robert J., Robert Milligan, William Lesser, William Magrath and Dale Bauman. 1984. Biotechnology and the Dairy Industry: Production Costs and Commercial Potential of the Bovine Growth Hormone. Cornell University Center for Biotechnology, A. E. Research 84-82. Ithaca, New York.

Martin, Daniel L. August 1985. Commercial Feasibility and Impact of Embryo Transfer Technology on the Dairy Industry: A Case Study. Unpublished M.S. Thesis, Department of Agricultural Economics, Texas A&M University, College Station.

Office of Technology Assessment. March 1986. Technology, Public Policy and the Changing Structure of American Agriculture. U.S. Congress, OTA-F-285. Washington, D.C.

Richardson, James W. and Clair J. Nixon. 1986. Description of FLIPSIM V: A General Firm Level Policy Simulation Model. Texas Agricultural Experiment Station, Bulletin B-1528. College Station.

Yonkers, Robert D., J. W. Richardson, R. D. Knutson, and B. M. Buxton. July 1986. Accomplishing adjustment in the dairy industry during technological change: the case of bovine growth hormone. Paper read at the annual meeting of the American Agricultural Economics Association. July 1986. Reno, Nevada.

7. The Social Impacts of Bovine Somatotropin: Emerging Issues

Bovine somatotropin (BST) is one of the most controversial technologies developed within the land-grant system, even though other technologies (e.g., hybrid corn) have had more substantial and more dislocating socioeconomic impacts. Nonetheless, projections of the potential socioeconomic and environmental impacts of BST have generated much controversy. It is important, therefore, to step outside the immediate realm of BST and its impacts and look at the broader implications of the controversy. We suggest that, in a broader context, BST raises a number of important issues that must be addressed in this emerging era of agricultural biotechnology. [1]

The development and deployment of BST are occurring, and will continue to occur, in an unprecedented milieu of changing land-grant politics, farm structure, farmer adoption behavior and management decision-making, and international division of labor in food production. In this chapter we lay out some of the considerations that are important in assessing the future impacts of BST.[2]

THE DEVELOPMENT OF BST
AND THE BST CONTROVERSY

BST, like a good many other technologies such as hybrid corn, was not developed solely within the land-grant system. Historically, the most crucial break-through in BST technology was recombinant DNA research

137

by Genentech. Following previous human growth hormone research, this resulted in the insertion of the somatotropin gene in a recombinant microorganism. BST, once a scarce substance, could now be produced in a factory and made available in unprecedented quantities for experimental research. Land-grant scientists -- most notably, D. E. Bauman of Cornell University -- ultimately did crucial research on the biochemistry of and administration and management procedures for BST. In the early 1980s, following the publication of seminal papers by Bauman and colleagues (e.g., Bauman et al. 1982), it became known that experimental BST results indicated that milk production per cow could be increased by up to 40 percent during the period of the injection. This led to widespread estimates that BST could permit a 25-30 percent increase in milk production per cow under farm conditions.

In large part because BST is essentially a bio-technology -- and potentially the first important commercial agricultural technology developed through biotechnology -- BST became the object of an unprecedented amount of _ex ante_ socioeconomic assessment research. The two earliest and most notable studies in this regard were by Kalter and colleagues at Cornell University (Kalter et al. 1985; Kalter 1985; Kalter and Milligan 1986; Kalter and Tauer 1987) and by the U.S. Office of Technology Assessment (OTA) in 1986.

The methods and "stylized facts" from these _ex ante_ studies were roughly as follows.[3] These studies assumed large revenue-over-cost returns from invest-ments in BST, based on the experimental data (and their "discounting" to farm-level conditions) referred to earlier. The Kalter group at Cornell collected data from a survey (see Kalter et al. 1985, for details) in which farmers were asked what their adoption plans would be given a scenario about BST costs, additional milk yield, etc. OTA (1986) collected data from panels of experts in the agricultural and social sciences in which, through a "consensor" technique, consensus was arrived at in terms of the configuration of the adoption curve, capital intensity of the technology, and so on.

Both studies projected very rapid adoption of BST -- roughly 80 percent within five years of commercial introduction -- a pace of adoption qualitatively different from previous dairy sector technologies (such as bulk milk tanks and artificial insemination), although OTA (1986) projected a slower pace of adoption. Both studies also projected substantial declines in the number of dairy farms and dairy farmers attributable to BST -- roughly 30 percent -- over a short period of time (within five years) (Magrath and Tauer 1986). There is apparently little debate on this point, since a substantial increase in milk production per cow -- should it ultimately be 10, 15, or 25 percent or more necessarily results in a proportionate, if not disproportionate, decrease in the number of dairy cows, and hence dairy farmers, assuming no major expansion of demand for dairy commodities.

The most controversial aspect of these _ex ante_ BST studies has been the distributional impact of BST among dairy farmers. OTA has taken the position that the adoption of BST will lead to disproportionate benefits among large dairy farmers and result in a disproportionate decrease in the number of small dairy farms (1986). The principal argument in the OTA assessment is that small operators will be at a disadvantage in using BST technology effectively because of their lack of access to information and their inferior financial situation relative to large operators. OTA has thus projected that large dairy farmers will be much more likely than their moderate- and small-sized counterparts to use BST technology; accordingly, large farmers would be expected to reap the lion's share of the benefits from the technology.

The Kalter group has focused less specifically on the distributional impacts of BST technology. Their work in this regard was largely limited to calculating the relationship between herd size and the timing of the first expected BST trial (see, e.g., Kalter et al. 1985). Kalter (1985) has nonetheless argued on the basis of results from a previous publication (Kalter et al. 1984) that:

> At the farm level there will be clear winners and losers. . . . [S]uperior farm managers often with large farm holdings and a secure financial position . . . will be the first to take advantage of the new technological innovations . . . and, therefore, [will be] the survivors of the next agricultural revolution. . . .[A]s in any time of technological change, early innovators will be more likely to survive. In addition to the financial benefit that early innovation provides,

it also allows an operator to move up the learning curve for the new technology before price begins to decline and additional productivity becomes necessary for survival. . . . [T]he introduction of biotechnology will probably accelerate the trend toward fewer and larger farms, and the structure of agriculture will tend more and more toward specialization. Only larger operations can develop a division of labor enabling the type of management practices being suggested here. Only specialization permits the focus necessary for superior management.

The position of Kalter and colleagues on this matter is less clear cut than this quote would suggest, however. Tangley (1986) noted that "Kalter emphasizes that bGH can help both small and large farms cut costs by making them more efficient. Moreover, he says, the trend toward fewer and larger dairy farmers is happening regardless of bGH."

THE CONTEXT OF THE BST CONTROVERSY

As noted earlier, BST has been a controversial technology well in advance of its commercial introduction. The controversy over BST erupted for several reasons. First, farm and public interest groups -- most notably the alliance between Jeremy Rifkin's Foundation on Economic Trends, the Audubon Society, and the Wisconsin Family Farm Defense Fund -- became aware of the ex ante impact assessment findings referred to earlier. These groups in particular as well as others and politicians, became critical of the land-grant universities and the federal government for devoting research funding and staff to developing a technology that would displace so many dairy farmers. Second, the prospect of a rapid increase in aggregate milk production and a decrease in milk prices was occurring at a time of farm crisis, particularly a crisis in general farm and dairy commodity policy. For example, net removals under the price support program in 1985 were about 10 percent of the aggregate level of production. This situation has changed significantly during the past several months as a result of the dairy herd buy-out program, but this is no doubt a temporary solution to a chronic surplus problem. BST promised to completely overwhelm the ability of the federal dairy commodity program to support prices above the average cost of production.

Another factor that intensified the BST conflict was the posture taken by the major land-grant universities, principally the University of Wisconsin and Cornell. This posture was largely a full-scale defense of the technology -- the assertion of the "scale neutrality" of BST and of the importance of BST in enabling the dairy industry to compete with producers of other beverages and foods and with foreign producers. This defense has at times also taken the form of discounting early results about how dramatic the BST impact would be on milk production per cow, aggregate production, and farm numbers.[4] Another posture has been to jump on the anti-commodity program bandwagon -- to say that the real problem is federal commodity program intervention, which has kept marginal dairy farmers in business too long, and which will interfere with adjustments to BST technology in the future. Many of the critics of BST have found these to be inadequate -- even insensitive -- explanations of the land-grant role. There has also been cynicism about the land-grant universities asserting the scale neutrality of BST when there is evidence to the contrary.

One of the intriguing dimensions of the land-grant response to the BST controversy has been their decision not to "get off the hook" by making the empirically tenable case that the recombinant somatotropin micro-organism (the key technical breakthrough in BST tech-nology) was not developed in the experiment station system. Indeed, most if not all land-grant universi-ties could have argued, with considerable justifica-tion, that their research was substantially devoted to enabling farmers in the state to use BST most effec-tively, should BST eventually be approved by FDA for commercial introduction. That this road was not taken, however, appears to have been due to the strong commitment of the major land-grant universities to biotechnology.

BST was widely hailed, during the early years of land-grant biotechnology expansion, as an example of an effective biotechnology that could revolutionize agriculture. It was a convenient example of a biotechnology that would almost certainly be successful and an example to state government officials that biotechnology will yield results. As a result, the land-grant universities felt it unwise to disassociate themselves from BST technology because it might have cast doubt on the wisdom of state governments supporting land-grant biotechnology programs as a whole.

SOME BROADER ISSUES IN THE BST CONTROVERSY

The Nature of Ex Ante Impact Assessment

Ex ante impact assessment is a relatively new area of work in both the rural social sciences and the social sciences.[5] In general, the approach dates from the early 1970s when environmental impact assessment and technology assessment generated a great deal of attention. The area of work has not been common in the land-grant system. It is probably fair to say that more ex ante assessment research is now being directed to the broad area of biotechnology than in the entire history of the land-grant system (and also over the entire repertoire of the technologies it has heretofore developed).

Ex ante assessment is a multifaceted phenomenon. One characteristic of this type of work is that it is difficult to do well. The future is inherently difficult to predict. Also, no one discipline has a monopoly on the skills necessary to project what typically are multifaceted and highly variegated effects of a technology over time. For example, a technology such as BST will not only have important aggregate economic impacts that economists are in the best position to assess, but quite likely distributional, political, and ideological impacts as well, which have not always been treated satisfactorily by economists (see Browne 1986, for a particularly useful assessment of the political and ideological nature of conflicts over BST). Unfortunately, it is still rare for social scientists from several disciplines to cooperate in ex ante assessment of emerging agricultural technologies in order to bring more skills and perspectives to bear on a common research problem. The OTA study is a notable exception. Still, agricultural economists predominated in its analytical phases.

A second characteristic of impact assessment of

new technologies is that, despite the difficulty of doing this work well, it tends not to be sophisticated work in a disciplinary sense, particularly after early assessments have been done and the methods have become routinized. In fact, it would be desirable for ex ante assessment to become a routine procedure so that results could be produced in a timely way by applying straightforward, routine methods. For example, the research necessary to compile a partial equilibrium or linear programming model of the dairy sector suitable for assessing the economic impacts of a technology such as BST would be of interest to disciplinary agricultural economists, but repeated iterations of the model to assess many new dairy technologies probably would not. Thus, to the degree that routinization occurs and our confidence in the quality of assessments increases, ex ante assessment would cease to be of great interest to the research community.

Third, ex ante assessments can be easily dismissed because of the unpredictability of the future. Also any particular assessment will have to focus on a limited number of assessment criteria, and hence will ignore some that some observers regard as important. In addition, impact assessment will necessarily have keenly interested external audiences -- those who stand to gain or benefit from the technology -- which may or may not be happy with the results of the research.

Collecting Ex Ante Assessment Data

The most valid primary social data are those that are being collected right now. People are directly or indirectly the data sources. If asked today what they have done or intend to do today, they can generally give more valid responses than if asked what they did 10 years ago or what they intend to do five years into the future.[6] Herein lies one of the central methodological problems of ex ante assessment.

Most impact assessment research is implicitly or explicitly based on calculating or making assumptions about the configuration of an "adoption curve." The two most important and interrelated parameters of an adoption curve are the percentage of farmers or other decision makers who adopt the technology over a given period of time and the rate of adoption. Ideally adoption curve data is available for various subcategories of farmers so that judgments about the differential costs, benefits, and impacts of the technology can be made.

Two basic methods have been employed to collect such data. Kalter and colleagues have used a mail survey approach in which farmers, after reading a fictional advertisement in Hoard's Dairyman and a fictional cooperative extension fact sheet, were asked a number of questions about their expected timing of adoption. While a creative way of collecting data, the problems with this approach are highlighted by the fact that the mail survey response rate was 13 percent, well below the typical rate of 65 to 70 percent in mail surveys among the general public.

OTA, as noted earlier, relied on the judgments of panels of experts in conjunction with a "consensor" procedure. These experts, while presumably having the benefit of years of observation about how dairy farmers adopt technologies of varying types, may also have certain biases -- for example, about superior management abilities and technology adoption records of large operators over small ones.

It is, in fact, surprising how much difference there is in the (implicit) adoption curves for farmers of different sizes estimated with these two methods. Kalter et al. (1985) reported a fairly modest effect of herd size on adoption rate. Farmers who indicated they would adopt BST within one year of commercial introduction had a mean herd size of 72.4 cows, those who would adopt within 1-5 years averaged 70.1 cows, and those who indicated they would adopt 5 years after commercial introduction (or never) averaged 49.5 cow herds. While OTA did not, to our knowledge, report specific data on adoption of BST by size of farm, data reported on the percent of farmers expected to adopt one of a larger set of "animal-biological" technologies (OTA 1986) indicate sharp disparities in adoption by gross sales class. The adoption rates varied from 10-20 percent for farms with less than $20,000 in gross annual sales to 80-90 percent for farms in the greater than $500,000 sales class.

One particular stylized fact about BST -- the extremely rapid rate of adoption -- may also be an artifact of methodology. In this case, both the Kalter group and OTA projected comparably rapid rates of adoption, giving the impression that confidence in this generalization is warranted because of similar results obtained with different methods. It is quite possible, however, that both methods may have overestimated the rate of response, albeit for different reasons.

When survey methodology is used, several considerations must be taken into account. One is that responses to questionnaires are often highly affected by the wording of questions and by the cues that the authors of questionnaires give to the respondents. If, for example, respondents are given hypothetical fact sheets and advertisements that paint a rosy picture of a technology, adoption rates will be biased upwards. Also, attentiveness to new technology is probably a matter of "social desirability" to farmers -- that is, farmers will be reluctant to admit that they are potential laggards or skeptical about adopting new technologies. Our guess is that "social desirability response set" may have contributed to the estimates of rapid adoption from survey techniques.

For example, there is evidence that BST has less effect on milk production in higher temperature zones. [7] Yet when the same hypothetical fact sheet and advertisements used in the New York Dairy survey were given to dairy farmers in the warmer states of Georgia, Florida, Mississippi, and Alabama, the anticipated adoption rates had the same upward bias.[8] Had the southern dairy farmers been given region-specific information in their fact sheets informing them of BST's sensitivity to temperature gradient, the ex ante adoption rates in the Southeast would have almost certainly been lower.

Discrepancies of this magnitude as a result of methodology are simply not acceptable if ex ante assessment is to be credible. Which way to turn? It is our feeling that a very substantial modification of the Kalter et al. approach will be most viable, since the OTA approach (assembling a wide range of experts across the country) will not be practical in most research. First, improve questionnaire design to enable far higher response rates in the 65-70 percent range (Dillman 1978). Second, survey studies should utilize "internal experiments" -- e.g., varying the degree of enthusiasm expressed in hypothetical literature or altering question wordings -- to measure how sensitive the results are to design procedures. Third, more attention will be needed to place adoption of BST and other technologies in the context of previous adoption and management decisions by farmers.[9] Fourth, mail or telephone survey results should ideally be complemented with indepth interviews to determine more precisely why farmers would or would not adopt a technology.

Finally, and most crucially with regard to the prospective impacts of BST, it may well be that the methodologies used to estimate the configuration of adoption curves have led to some exaggeration of the rate of adoption. And since the dislocations that will result from BST are, in part, a function of the rate of adoption, these estimates may have contributed to an exaggeration of these dislocations.

TECHNICAL CHANGE AND THE THREE AGRICULTURES

Most of our theory and empirical research on farmer adoption of new technology has been rooted in a particular period of U.S. agricultural development -- a period during the 1950s and 1960s in which there was only modest differentiation between small and large farms and when most farms absorbed between one and two person-years of labor equivalent. The structure of American agriculture has now begun to diverge markedly from this post-World War II pattern. There has been a slow but steady trend toward a more dualistic farm structure that has three major components. The first group includes 25,000 very large farms -- a little over 0.1 percent of total U.S. farms -- most of which depend primarily on hired labor. This handful of very large farms now accounts for nearly 35 percent of annual gross sales. The second includes roughly 1.7 million small or "subfamily," primarily part-time farms (with gross annual sales of less than $40,000), which account for about 70 percent of the U.S. number. Very large and very small farms have both been increasing in numbers over the past 15 or so years. The third component of the emergent dualistic structure is the "disappearing middle" of medium-sized, primarily full-time, family-type farms. The "middle" of medium-sized, full-time family farms is said to be slowly disappearing because they enjoy neither the "advantages of bigness" (scale economies, volume discounts on inputs) nor the "advantages of smallness" (high levels of off-farm income).

It is quite possible that if this dualistic tra-

jectory of farm structural change continues, the three
major types of farms may increasingly correspond with
three very different sets of criteria for adoption
decision making. Sociologists and anthropologists, for
example, have long had evidence that family farmers
make decisions in ways that do not correspond with
postulates such as those in capital budgeting models
(see, for example, Bennett 1982; Bennett and Kanel
1983). This is because farmers' decision-making criteria tend to focus more on factors such as holding onto
the family farm instead of achieving the average rate
of profit in the economy at large, which may not be
particularly important.

The major sociological perspective on technological decision making -- the so-called adoption-
diffusion perspective -- is not obviously superior to
the economists' assumptions that a particular decision-
making calculus pertains to all farmers. The adoption-
diffusion perspective has hypothesized that farmer
adoption of technology is determined by a set of
"traditional-modern" social-psychological orientations;
farmers with "modern" orientations are more likely to
adopt new technologies at an early point than farmers
with more "traditional" orientations. Value-
orientations that could be arrayed on a traditional-
modern continuum were influential in shaping
technological adoption decisions during the first few
decades after World War II. However, now that farmers
have increasingly greater educational backgrounds and
are more closely integrated into the larger society
through mass communication (which has reduced the role
of traditional values as sociologists have
conceptualized them), there is reason to believe that
these factors are no longer very important in
technological decision making. They are probably much
less important than the more concrete interests of
different types of farmers that we will discuss
shortly.

There is also evidence that there is a growing
class of farmers -- the 25,000 or so very large farmers
referred to earlier -- for whom the assumptions of
capital budgeting and related models are appropriate
(Sonka 1983). These farmers are essentially portfolio
managers -- entrepreneurs who currently hold farm
assets, but who would not be reluctant to liquidate
these farm assets if returns would be higher in other
businesses --as opposed to the family farmers that

noneconomists such as Bennett (1982) and Rogers (1987) have based their research.

Agricultural census data on smaller, part-time farmers demonstrate that these farmers' decision-making criteria are not based primarily on maximizing returns to equity capital. For example, the 1978 and the 1982 census revealed that the average net farm income of operators with gross annual sales of less than $40,000 (1982 constant dollars) was negative. For many of these operators, farming is pursued to hold onto landed assets for lifestyle or for other "noneconomic" reasons. Perhaps more importantly, many (though by no means all) of these farmers can remain in business with no returns or modest long-term losses because of their off-farm income, which averaged nearly $20,000 per household in 1982 (OTA 1986), and which is known to be substantially higher on the smaller farms (those with annual sales less than $20,000). Small farmers also tend to have very low debt loads, which insulates them from adverse price changes.

These considerations are important for several reasons. First, to confidently project that a given percent of a particular type of farm operator will adopt a technology such as BST, an assessment should ideally have a more theoretical basis in farm structure and decision making. Second, the preceding hypotheses about differences among farm types in decision-making behavior have obvious implications for BST adoption. One of these is that small, part-time farm operators may not be particularly vulnerable to being forced out of farming, regardless of whether or not they adopt BST. Medium-sized, full-time family dairy operations may be the most vulnerable to the dislocating impacts of BST, though most would not be likely to quit farming entirely (even if they decide to quit dairying). We might expect these operators to shift enterprises -- particularly to less labor-intensive enterprises that will free up labor that can be allocated in off-farm labor markets.

Perhaps the real wild card in the BST adoption experience will be how larger -- portfolio manager-type farm -- operators make their decisions. OTA (1986) has projected that these farmers, because of their high returns to investment, will be best able to weather the shocks that result from BST-induced milk price de-clines. But will farmers who are essentially portfolio managers (e.g., industrial dairy operators in the Sun-

belt states) be content to merely stay in business?
Will the effects of BST cause a substantial number of
them to liquidate their dairying assets and invest
their capital elsewhere? If BST leads to a relatively
long-term period of depressed milk prices and low returns, even to superior managers making effective use
of BST, would we witness a renaissance of family dairy
farming and a demise of industrial dairying? We lack
answers because we have not adequately theorized farmer
decision making in an increasingly differentiated and
dualistic farm structure.

BIOTECHNOLOGY AND THE LAND
GRANT POLITICAL BASE

It may sound like an irresponsible statement, but
it has long been known that the rank and file of
farmers tends not to benefit, and often loses, because
of technical change (even though society as a whole is
arguably better off). This is because of the tendency
for "continuous technological change . . . [to result]
in excess resources in agriculture, which generate
lower average rates of return than what would exist
under no technological change" (Tauer 1987).   This
point has been made by well known social scientists
(e.g., Schultz 1977; Cochrane 1979).
Accordingly, farmers as a whole have historically
never been strongly supportive of agricultural
research. Many farm groups, in fact, resisted the Hatch
Act and later pieces of agricultural research
legislation.   Farmers, in part because of their
tendency to be ambivalent about agricultural research
and technology, have never been enthusiastic supporters
of agricultural research at the federal level.   The
principal reason that the land-grant system has grown
to its current level, however, is that farmer
ambivalence about research at the federal level has
been substantially mitigated by funding research at the
state level.   Farmers at the state level --
particularly early adopters in a position to receive
innovators rents -- can be induced to support research
that helps them compete with farmers in other states.
The land-grant system thus has come to be dominated by
state appropriations.   The raison d'etre of the land-
grant system is to conduct applied, locally-adapted

research that is disproportionately appropriable by farmers in a particular state (Buttel and Busch 1988).

BST is a biotechnology. And biotechnology promises to alter the political demand structure of agricultural research for several reasons. First, biotechnology research is generally not applied or problem-solving research with priorities closely shaped by the expressed technical needs of farmers in the state. Second, biotechnologies -- at least the early ones such as BST -- are usually not, at least by choice, locally adapted technologies.[10] In particular, BST is, generally speaking, just as applicable to California dairying as it is to the New York and Wisconsin dairy industries.[11] Third, the direct or ultimate clients of biotechnology research are primarily out-of-state (if not foreign) corporations, rather than state agribusinesses. [12]

BST has probably been controversial primarily because it epitomizes the generic technologies -- technologies applicable over large geographical areas -- currently sought in biotechnology research. If BST technology developed at Cornell or Wisconsin was disproportionately implemented by New York and Wisconsin dairy farmers, there would not have been the struggles over this technology that we have witnessed in recent years.

Will the BST controversy and similar ones to follow -- perhaps over porcine growth hormone, because of its adverse impacts on grain producers (Kalter and Milligan 1986) -- begin to undermine farmer support for land-grant research, and hence the state-level funding base of agricultural research? There is no way to know at this point. And, practically and scientifically speaking, there is no going back on the commitment to biotechnology in the land-grant system. But we suspect it will be important to the future of the land-grant system for administrators to recognize that fundamental biotechnology research challenges the political and funding underpinnings of the experiment station system. Strategies will need to be developed to shore up the commitment to serving state-level farmer groups in the new age of agricultural biotechnology.

PRODUCT SUBSTITUTION AND THE DAIRY INDUSTRY

The nature of biotechnology is that it provides an essentially common technical basis to a wide variety of products -- pharmaceuticals, plant agriculture, animal agriculture, chemicals, pollution control, food processing -- that formerly were quite separate technically. Accordingly, one cannot assess the impact of a technology such as dairying on a sector by examining that sector in a vacuum. In particular, biotechnology raises potentially profound possibilities for product substitutions of various sorts. Industrially-developed product substitutions in agriculture and dairying are, of course, by no means new. For example, beginning at the turn of the century, milk fats (butter) were replaced by vegetable oils and fats (margarine). But biotechnology promises to accelerate these product substitutions.

Many biotechnology-related examples have emerged in world agriculture over the past decade -- for example, high-fructose corn syrup and aspartame for sugar, tissue-cultured palm oil for coconut oil, as well as many examples of industrially-produced flavors, fragrances, and other compounds substituting for chemicals formerly extracted from plants. Others are likely, e.g. corn for potato starch, single-cell protein for soybeans and other protein sources, and enzymatic production of feed additives to transform protein-deficient grains into nutritionally-rounded animal feed (van den Doel and Junne 1986; Kenney, Buttel, and Kloppenburg, Jr. 1984).

BST is not the only biotechnology that will affect dairying. One of the major threats to the dairy industry could be the application of biotechnology to changing the properties of vegetable proteins (rape seed, perhaps) to enable them to be a suitable substitute for casein. It has been estimated, for example, that if all the cheese made in the European Economic Community were derived from "vegetable casein," about 3.2 million cows (about 13 percent of the European dairy herd) would no longer be needed (Bijman 1986; see Goodman, Sorj, and Wilkinson 1987 for a comprehensive discussion of substitution effects of biotechnology). It should thus be recognized that the socioeconomic impact of biotechnology on the U.S. dairy industry is not merely a matter of whether U.S. dairy farmers and the dairy industry will benefit from BST.

152

CONCLUDING COMMENTS

In conclusion we have some final observations on BST in the context of impact assessment, land-grant policy, and the future structure of the international division of labor in food production. First, it is important to stress the importance of ex ante assessment of emerging agricultural technologies, despite its methodological challenges within the land-grant system. Some might prefer the good old days when scientists could proceed with research projects without social scientists or the public looking over their shoulders. But those days are probably long gone, as Ruttan (1987) has cogently argued. In this new era of public agricultural research, which is defined by stagnation of public funding of research, an increased emphasis on basic or fundamental research, closer industry-university relationships, a more dualistic farm structure, farm financial stress, and the growing sophistication of farmer groups about the kinds of technologies they prefer, the information that can be made available through ex ante assessments will be increasingly essential for research administrators and scientists.

Second, as sociologists we have great qualms about the frequent use of the notion of "scale neutrality" in describing BST. Scale neutrality is a concept that depicts the likely distributional consequences of a technology (i.e., either the distributional impacts of a technology across different classes of farmers or sizes of farms, or distributional impacts vis-a-vis changes in shares of returns to land, labor, and capital). We believe, instead, that BST and probably most other future agricultural biotechnologies are "divisible" inputs (i.e., BST is not an input that requires "lumpy" investments such as large machinery). Divisibility is, in general, a prerequisite for, but does not guarantee, scale neutrality as we have defined it (i.e., the neutrality of outcome). As we know from long debates over the impacts of the Green Revolution in the developing countries, scale neutrality depends, in part, on the characteristics of the technology (e.g., "divisibility" vs. "lumpiness," capital-intensity, labor-intensity, effects on aggregate output) and also on the characteristics of the socioeconomic context within which the technology is deployed (e.g., see Lipton with Longhurst 1985).

Third, we need to develop a more fine-grained view of agricultural biotechnologies to put BST into a broad context. As noted earlier, there are many types of agricultural and industrial biotechnologies. Moreover, these biotechnologies are being developed at different paces due to a variety of factors (e.g., technical difficulty, market potential, the allocation of public basic research funding across technical areas). Barker (1986) has pointed out, for example, that the most rapid biotechnology advances are being made in (1) the animal sectors, (2) the area of industrial biotechnology substitutes for primary agricultural commodities, and (3) high-value horticultural crops. The slowest pace of advance is in the cereal grains.

The great unevenness in the timing of biotechnology advances might have major effects on the structure of the international division of labor in food production. Over the past two decades, the advanced industrial countries have become increasingly specialized in the capital-intensive production of food and feed grains, while livestock production (other than dairying) and production of high-value horticultural crops has slowly but surely shifted toward the Third World.[13] That is, the United States and other advanced countries have become specialized in the types of commodities for which major commercial biotechnology advances will be slow in coming, while many of the currently more technologically dynamic sectors have tended to shift toward the developing world. Will the uneven timing of biotechnology advances cause a reversal of this division of labor, which results in a higher proportion of world livestock and high-value horticultural crop production returning to the advanced countries, while cereal grain production becomes increasingly relegated to the Third World? This question cannot be adequately addressed at this point. It is, however, illustrative of the very broad context in which biotechnology might operate in the future and of how the effects of biotechnology will need to be considered at a global level. For biotechnology is a global technology that will have global, as well as highly localized, impacts.

Finally, it will be important to recognize that BST is a politicized agricultural technology _sui generis_. This technology will have some distinctive features that will render results from _ex ante_ (and ultimately _ex post_) socioeconomic research on BST only partially generalizable to other agricultural biotechnologies. Perhaps the most distinctive aspect of BST may be its politicized character. One aspect of its politicization is that virtually all dairy farmers will know a great deal about BST and its pros and cons well before it is commercially available (Heuth and Just 1987). Such widespread farmer knowledge on a technology prior to commercialization is quite rare and may serve to hasten its adoption by farmers.

In a sense, then, research indicating that BST will be diffused very rapidly may inadvertently become a self-fulfilling prophecy. Another feature of the politicization of BST is that we may expect farmers to adopt or not adopt the technology based on their political ideologies and on their orientations toward agribusiness and the postures that the opposing sides have taken in the BST controversy.[14] Other distinctive aspects of BST (relative to most other agricultural technologies) include its potential to increase aggregate output rapidly (i.e., result in rapid shifts in the supply curve), the fact that the results from its use by farmers are immediate and highly discernible, and the extent to which the benefits of the technology will be distributed according to farmers' degree of specialization in dairying and their management abilities. In sum, we should be cautious in assuming that extensive research on BST will allow us to generalize about the future impacts of biotechnologies as a whole.

NOTES

1. Paper prepared for the National Invitational Bovine Somatotropin Workshop, sponsored by the Extension Service, U.S. Department of Agriculture, St. Louis, September 22, 1987. The authors would like to thank Loren Tauer, Robert Kalter, and R. David Smith for their helpful comments on an earlier draft. See, for example, Kloppenburg, Jr. (1984) for an examination of hybrid corn technology in a qualitative, ex post impact assessment framework.

2. This chapter focuses only on the prospective socioeconomic impacts of BST, and thus does not consider two other matters that have been at issue in debates over this technology: (1) human health and

safety aspects of BST-produced milk, and (2) herd health impacts of BST. It should also be noted that we have not done extensive primary *ex ante* impact assessment research on BST. Our interest in BST has been part of a larger concern with the nature of biotechnology research in agriculture (see, for example, Buttel 1986).

3. By "stylized facts," we mean frequently cited results of studies that are themselves not often read. Thus, the stylized facts from BST impact assessment studies are sometimes different from those the authors of the studies have written.

4. This argument, as we have suggested elsewhere (e.g., Buttel 1986), is probably accurate (see also, for example, Tauer 1987). However, this argument, is clumsy because it is generally inconsistent with the *ex ante* impact assessment results that were published at the time.

5. It could, of course, be argued that *ex ante* impact assessments have been conducted for many years. Goldschmidt's (1947) analysis of the likely impacts of Bureau of Reclamation investments in irrigation in California is a good example from anthropology. In the 1930s and 1940s many rural sociologists did research on the future impacts of mechanization in Southern agriculture. Economists, by developing macroeconomic and microeconomic models and doing future projections, have done parallel work for a long time. But prior to the 1970s there was not a substantial research community in any of the rural social sciences focused on assessing the socioeconomic impacts of emerging technologies.

6. There are, of course, clear caveats to this generalization. Sociological research (e.g., Dillman 1978) has demonstrated that a personal interview or telephone or mail survey response is a highly structured situation of social interaction. For example, respondents will have a tendency to provide responses that they feel the interviewers or survey authors want or expect. In surveys of the general public, respondents tend to be reluctant to provide answers that reflect racist attitudes. As will be discussed below, it is quite possible that respondents in surveys about agricultural technology adoption will tend not to admit that they are inattentive to new technologies.

7. Conversation with Dr. David Kronfeld, School of Veterinary Medicine, University of Pennsylvania, April 3, 1987.

8.   Phone conversation with Dr. Henry Kinnucan,
Agricultural Economist, Auburn University, Alabama,
April 18, 1987.

9.   For example, one might determine whether prospective early adopters of BST tend to be participants
in DHIA programs, users of artificial insemination,
adopters of conservation tillage practices, and so on.
Knowing the kinds of technologies adopted in the past
by prospective early and not-so-early BST adopters
would help to provide a better understanding of the
motivations and adoption considerations of various
types of farmers.

10.   It should be noted, however, that there is
some debate about whether the pattern of structural
change in U.S. agriculture is clearly dualistic -- in
particular, about whether there has been a "disappearing middle" (see, e.g., Buttel and LaRamee 1987).

11. Some agricultural biotechnology products
(e.g., recombinant free-living, N-fixing soil microorganisms, minor crops grown in small areas) may be
"locally-adapted" in a sense. This is not, however, by
design -- i.e., to meet the technical needs of farmers
in a state. Moreover, private firms will have a strong
preference for relatively generic technologies that can
be marketed over large geographic areas.

12.   A few states, primarily on the East and West
Coasts, have a substantial number of small start-up
biotechnology companies. Most of these start-up firms,
however, have major relationships -- especially stock
ownership -- with the larger multinational
biotechnology firms (for example, see Buttel et al.
1986, for a discussion of these patterns within the
context of biotechnology research in the land-grant
system).

13.   The principal reason for this pattern is that
labor-intensive horticultural production and land-
extensive livestock production have moved toward zones
with cheaper labor and land (Sanderson 1985).   Also,
the industrial-temperate zones are most conducive to
food and feedgrain production, most research on these
crops is conducted there, and these crops have been
heavily subsidized by industrial countries'
governments.

14.   Previous research on farmer adoption of antibiotics in animal feeds, for example, has indicated
that nonadopting farmers tend to be politically lib-
eral, to have reservations about agribusiness, and to

disbelieve industry's positions (e.g., that protracted subtherapeutic use of antibiotics in animal feed will lead to resistance in animals) in the scientific debate that has accompanied this technology (Gillespie, Jr. and Buttel 1987).

REFERENCES

Barker, R. 1986. Impact of prospective new technologies on crop productivity: implications for domestic and world agriculture. Paper presented at the Conference on Technology and Agricultural Policy, December, National Academy of Sciences, Washington, DC.

Bauman, D. E., et al. 1982. Effect of recombinantly-derived bovine growth hormone on lactational performance of high yielding dairy cows. _Journal of Dairy Science_ 65 (Supplement 1):121.

Bennett, J. W. 1982. _Of time and enterprise_. Minneapolis: University of Minnesota Press.

Bennett, J. W., and Kanel, D. 1983. Agricultural economics and economic anthropology: confrontation and accommodation. In _Economic anthropology_, ed. S. Ortiz. Lanham, MD: University Press of America.

Bijman, J. 1986. Substitution processes in European agriculture. _Trends in Biotechnology_ 4:91-92.

Browne, W. P. 1986. Bovine growth hormone and the politics of uncertainty: fear and loathing in a transitional agriculture. Manuscript, Department of Political Science, Central Michigan University.

Buttel, F. H. 1986. Agricultural technology and farm structural change: bovine growth hormone and beyond. _Agriculture and Human Values_ 3:88-98.

Buttel, F. H., and Busch, L.. 1988. The public agricultural research system at the crossroads. _Agricultural History_ 62:forthcoming.

Buttel, F. H., Kenney, M., Kloppenburg, Jr., J., and Smith, D. 1986. Industry-university relationships and the land-grant system. _Agricultural Administration_ 23:147-81.

Buttel, F. H., and LaRamee, P. 1987. The 'disappearing middle': a sociological perspective. Paper presented at the annual meeting of the Rural Sociological Society, August, in Madison, Wisconsin.

158

Cochrane, W. W.   1979. <u>The development of American
    agriculture</u>. Minneapolis: University of Minnesota
    Press.
Dillman, D. A. 1978. <u>Mail and telephone surveys</u>. New
    York: Wiley.
Gillespie, Jr., G. W. , and Buttel, F. H.   1987.
    Technological politicization and the ideological
    dimension of technological change: the case of anti-
    biotic animal feed additives. Paper presented at the
    annual meeting of the Rural Sociological Society,
    August, in Madison, Wisconsin .
Goldschmidt, W. 1947. <u>As you sow</u>. Glencoe, IL: Free
    Press.
Goodman, D., Sorj, and Wilkinson, J.   1987. <u>From
    farming to biotechnology</u>. Oxford: Basil Blackwell.
Heuth, D. L., and Just, R. E.   1987. Policy implica-
    tions of agricultural biotechnology. <u>American Jour-
    nal of Agricultural Economics</u> 69(May):426-31.
Kalter, R. J.   1985.   The new biotech agriculture:
    unforeseen economic consequences. <u>Issues in Science
    and Technology</u> 2:125-33.
Kalter, R. J., and Milligan, R. A.   1986.   Emerging
    agricultural technologies: economic and policy
    implications for animal production. Manuscript,
    Department of Agricultural Economics, Cornell Univer-
    sity.
Kalter, R. J., and Tauer, L. W.   1987. Potential
    economic impacts of agricultural biotechnology.
    <u>American Journal of Agricultural Economics</u> 69:420-25.
Kalter, R. J. et al.   1985.   Biotechnology and the
    dairy industry: production costs, commercial poten-
    tial, and the economic impact of the bovine growth
    hormone. A. E. Research 85-20. Ithaca: Department of
    Agricultural Economics, Cornell University.
Kalter, R. J., et al.   1984.   Biotechnology and the
    dairy industry: production costs and commercial
    potential of the bovine growth hormone. A. E. Re-
    search 84-22.   Ithaca: Department of Agricultural
    Economics, Cornell University.
Kenney, M., Buttel, F. H., and Kloppenburg, Jr., J.
    1984. Impact of industrial applications: socioeco-
    nomic impact of production dislocation. <u>ATAS Bulle-
    tin</u> (Advance Technology Alert System, UN Centre for
    Science and Technology for Development, special issue
    on Tissue Culture Technology and Development) 1:48-51.

Kloppenburg, Jr., J. 1984. The social impacts of biogenetic technology: past and future. In *Social Consequences and Challenges of new agricultural technologies*, eds. G. M. Berardi and C. C. Geisler. Boulder, CO: Westview Press.

Lipton, M., with Longhurst, R. 1985. *Modern varieties, international agricultural research, and the poor*. Washington, DC: World Bank.

Magrath, W. B., and Tauer, L. W. 1986. The economic impact of bGH on the New York State dairy sector: comparative static results. *Northeastern Journal of Agricultural and Resource Economics* 15:6-13.

Office of Technology Assessment (OTA). 1986. Technology, public policy, and the changing structure of American agriculture. Washington, DC: OTA.

Rogers, S. C. 1987. Mixing paradigms on mixed farming: anthropological and economic views of specialization in Illinois agriculture. In *Farm work and fieldwork*, ed. M. Chibnik. Ithaca: Cornell University Press.

Ruttan, V. W. 1987. Agricultural scientists as reluctant revolutionaries. *Choices* (Third Quarter):3.

Sanderson, S., ed. 1985. *The Americas in the new international division of labor*. New York: Holmes & Meier.

Schultz, T. W. 1977. Uneven prospects for gains from agricultural research related to economic policy. In *Resource allocation and productivity in national and international agricultural research*, Arndt et al. eds. Minneapolis: University of Minnesota Press.

Sonka, S. T. 1983. New approaches to farm and ranch management. In *Agriculture in the twenty-first century*, ed. J. W. Rosenblum. New York: Wiley-Interscience.

Tangley, L. 1986. Biotechnology on the farm: will new technology further disrupt our troubled agricultural communities? *BioScience* 36:590-93.

Tauer, L. W. 1987. Economic changes from the use of biotechnology in production agriculture. Paper presented at the annual meeting of the Iowa Academy of Science, April 24, at Grinnell College.

van den Doel, K., and Junne, G. 1986. Product substitution through biotechnology: impact on the third world. *Trends in Biotechnology* 4:88-90.

8. Biotechnology and the Future of Production of Meat and Other Livestock Products

Can biotechnology research and development help increase the efficiency of meat production and help maintain present abundance of food at relatively low prices? This paper argues that even small increases in biological efficiency of production may be difficult and limited, but small ruminants offer one avenue where production ceilings remain quite distant.

Common crops such as corn, soybeans, wheat, feed grains, milk, beef, pork, and poultry are nearing their peaks or limits in efficiency no matter what technology is developed. All of these products or crops require inputs of feed or fertilizer equal to the content of the products harvested. Sometimes a biological limit is reached depending on the capacity of the animal to take in more feed or the plant to transform fertilizer to heavier weight of seed or whatever the product is. Poultry for meat may have approached the ceiling already.

Efficiency in small ruminants depends more on the number of offspring per female than on the weight of individual offspring. The output is not feed dependent as the number of adult ewes may be reduced as the number of offspring increases, thus holding feed (generally natural forage) costs constant and increasing the number of market adults as well as lambs for meat. It has become increasingly clear that effective maintenance research is necessary to sustain the high yields now being obtained. Only small ruminants, which are also low cost, are below their ceiling in efficiency, possibly for a period as long as 50 years or more in the future. These possibilities have resulted from considerable past research, but whether they accelerate or even maintain the gains from present research may be questionable.

For example, "geeps" are brought about by injecting cells from a goat embryo and inserting them into a sheep embryo. The assumption that a geep would be more useful than sheep or goats just because we haven't had such an animal before is not reasonable. The hope for accelerating crossbreeding is mentioned, but full advantage of crossbreeding within species has already been taken. Furthermore, leaner carcasses from small ruminants are already being produced abundantly. The potential to increase efficiency of small ruminant meat animals through biotechnology selection research is so great that it will be difficult for any other technology to have a practical impact on farm production.

Davidson (1985) seems optimistic that biotechnology through genetic engineering, cloning, new biological processing techniques, and other scientific applications of molecular biology will lead to revolutionary change. He states, "For the first time in our history, all of the life sciences have a common basis in the language of molecular and cellular biology. The age of biotechnology is this revolutionary new perspective on the fundamental mechanisms of life and how they can be applied to our well being." He predicts a dramatic reduction in disease loss from isolation and cloning of genes that control immune systems and also from the recognition of genes associated with disease inheritance. On the other hand, past research has been effective in reducing losses from disease so that losses from predators, mismanagement, undernutrition, and lack of adaptation to environmental stresses are overshadowing livestock losses from disease, except as a secondary result.

Anderson (1986) discusses gene transfer techniques in animals and finds only two (microinjection and retroviruses) that have proven to be effective for inserting functional genes into intact animals. He predicts that improved techniques for genetic engineering will arise over the next few years. Rexroad (1986) outlines the requirements for using gene insertion in farm animals but concludes that predictions about the development of useful livestock are difficult to make. The 1985 Symposium on Genetic Engineering of Animals in Davis, California (Anderson 1986), led me to believe that while great potential exists, the realization of this potential may be many, many years in the future. In the meantime, genetic engineering, not single gene transfer, may lead to unprecedented gains in efficiency of meat production from ruminants. This will be accomplished by the detection of combinations of genes contributing to production efficiency and by combining innovative selection and reproductive techniques along with much more rapid turnover of generations.

MEAT PRODUCTION

Well over half of the meat produced in the United States is produced from grain fed to pigs and chickens. Beef accounts for most of the remainder with calves being raised on forage and then being fed largely grain for a considerable period before slaughter. Also grain is very important in the production of eggs and milk. In general, meat production has increased especially from poultry, due primarily to high efficiency and cheap grain. Likewise, milk production has increased due partly to cheap grain and partly to the government price support program. Both beef cattle and swine populations have declined in part because of low prices, but often because they could be liquidated to satisfy banker demands, possibly because of money lost on crop lands. The full effect of reducing beef cattle numbers is not evident yet (Economic Research Service 1986; 1986).

Production of corn, sorghum, and soybeans are being depressed by diverting land into soil conservation reserves, and by diversion to non-use. In addition, marginal land is being abandoned because support prices are inadequate for the farmer to make a living. There is also increasing demand to convert more corn to ethanol to help the domestic economy, to reduce the foreign payment deficit, and to reduce the need for imports of fossil fuel. Sooner or later the prices of grain and soybeans will probably rise at least slightly. This will tend to suppress production of meat, milk, and eggs, as well as feed exports. It will also likely result in higher prices to consumers. Non-feed costs for production and processing of all livestock products are continuously rising so that if farm and retail prices do not increase, production and consumption will be further suppressed. In general all livestock products are fully consumed. Consumption can only increase if production or imports increase.

Forages are generally the lowest cost feed for livestock, especially natural forage on land that cannot be utilized for other crops. There are hundreds of millions of acres of such land in the United States, much of which is unused, given up to scrub trees and brush, or underused by light or no grazing. However, with competition from cheap grain it may not be economical to use such land for food production, even though the land cost is low because of the high cost of utilizing the pasture or range for labor, fences, and predator losses. If and when the prices of grain and soybeans do increase because of reduction in surpluses, there will probably be a shift toward greater use of forage by livestock rather than a reduction in livestock numbers. Greater use of forages by livestock will be more certain if efficiency increases. If livestock and meat production could be increased, it should improve both health and standard of living.

RUMINANTS

Ruminants can live on forages alone. Thus meat and other animal products can be produced entirely from materials that cannnot be consumed by people. Cattle can be supported on the better pastures and ranges, but more than one year is required to reach market weight. Finishing on cheap grain in feed lots is the general practice. Small ruminants, sheep, and goats, can be marketed within one year from date of breeding and utilize the poorer ranges and pastures. It is generally economically advantageous and gives more complete utilization of natural forages if all three species are grazed together, but relatively large farms or ranches are required for this to be practical.

Small ruminants have advantages over cattle because of their earlier age of puberty, shorter gestation periods, much higher productivity, growth to market weight in 4 to 6 months, and the ability to produce high quality meat on forage alone and often on lands that cannot be used by cattle or for other purposes. However, the greatest advantage of small ruminants from an efficiency standpoint is that male and female generations can be turned more rapidly than for cattle.

Male offspring, selected on the productivity of their mothers, permit a new generation of sires each year. Females can be bred to have offspring at one year of age, and about half which wean offspring (superior for reproduction rate) can be saved for replacements, with the remainder leaving the breeding flock. Then, as the number of offspring increase, additional adults may be removed from the flock so that the number of parents plus offspring is kept constant, and thus, feed costs are also held constant. Both practices tend to also reduce generation length for females. This reduction will continue as the reproductive rate increases.

It has been shown at the U.S. Sheep Experiment Station, Dubois, Idaho, that the weight of lamb produced per female kept can increase at the rate of over two percent per year, even under highly extensive production. Furthermore, states in which farmers purchase more rams from the station have also had greater increases in lamb crops and net returns as compared with those that purchased fewer rams.

Generation length or age of the parents when the offspring are born is probably more important in genetic progress per year in selection of small ruminants than heritability or the selection differential because it can be reduced from about four years to one year, depending on management and reproductive techniques used. The other pertinent factors are heritability, which is generally fixed, and the selection differential, which can be changed by management within fairly narrow limits. Genetic progress per year is estimated by multiplying the selection differential by heritability and then dividing the resulting product by the generation length. Obviously, dividing by one gives much greater progress than dividing by four. Furthermore, if genetic progress is being made from selection, the youngest animals in years always have the highest genetic merit so that reducing the generation length insures that younger animals are favored in selection.

MEAT FROM SMALL RUMINANTS

Meat from small ruminants is a minor component of consumption in the United States. The low consumption of lamb is often used to claim that the demand for lamb

is low. The low consumption of lamb is not due to lack
of demand, but rather because production has been
drastically reduced by predator losses. Thus, there are
many places where lamb is not available even though
people there would like to eat it. This depression of
lamb production by predators has been going on for over
two decades so that many have grown up with no
opportunity to eat lamb. There has also been some
aversion to lamb, which developed from the serving of
low grade mutton to the armed services in World War II.
However, as lamb and goat meat production increases in
the future because of high efficiency of production and
because of the increased utilization of low value land,
there is no doubt that it will be fully consumed because of its high palatability, tenderness, and reasonable price.

Increased meat production from small ruminants
through biotechnology research offers the best hope of
increasing farm income and reducing the loss of farmers
in the future. In 1983-1985, net returns to earned
inputs would have been $24, $64, and $93 per acre (4
ewes) respectively from sheep. This would have given
an advantage of sheep in 1985 of $114 per acre over a
cow-calf operation (2/3 cow per acre), of $31 per acre
over soybeans, and of $27 per acre over corn.
Smallruminant meat crops with low costs, low capital,
and low energy requirements can slow land erosion,
utilize abandoned and non-crop land, and can be
produced in every rural county in the country.

Sheep and goats are a very minor source of food in
developed countries, but usually a major source of food
in developing countries. Each can produce high quality
products on forage alone, but they have difficulty in
competing with cattle, swine, poultry, and fish when
surplus feed is cheap. If and when feed prices return
a fair cost of living to producers, the retail prices
of beef, pork, poultry, and farm raised fish will
probably double or triple. Efficiency of meat
production from sheep has been declining due to
predator losses, but it will probably increase by
about one-half percent per year if predator losses can
be reduced to a low level by use of single lethal dose
baits. If increased biotechnology research and development is initiated, the increase in efficiency or net
return per acre of the sheep industry might exceed five
percent per year.

BIOTECHNOLOGY RESEARCH AND DEVELOPMENT

Biotechnology research and development consists of combining genetic and reproductive techniques to obtain a more rapid increase in production of weight of off-spring per female per year. At the same time, the techniques can be used to select aspects of efficiency that require costs and technology that the farmer cannot afford. Aspects other than reproductive efficiency include genetic improvement in feed efficiency, meat quality and flavor, resistance to disease and parasites, fertility of frozen semen, removal of seasonality of breeding, synchronization of ovulation,and embryo transfer. Genetic improvement in research nucleus flocks, within ecosystem areas, and including all important breeds would be passed on to farmers through sale of improved rams and later through sale of improved ram semen.

Biotechnology will be used to improve the overall quality as well as the weight of wool or mohair marketed per year. The approaches would basically include rapid selection for reproductive rate, including the weaning of heavy multiple offspring. Other traits including growth rate, feed efficiency, meat quality and flavor, fertility of frozen semen for artificial insemination (AI), ability to breed at any time of the year, and more often than once per year, and fiber weight and quality. These traits would be superimposed on selection for reproductive rate by use of a ram lamb or male kid index for each breed and ecosystem area based on the economic importance of each trait within a particular ecosystem area. Emphasis on weight of offspring weaned per female per year would give ample emphasis to milk production, mothering ability of females, and weaning weight.

Offspring would be weaned at about 4 to 5 months of age. They would be given supplemental feed through breeding, which should begin at 7 months of age so that their offspring would be born at about one year of age. Ram lambs would be fed individually from about 4 to 7 months of age to obtain information on feed efficiency. Ewe lambs would be fed so that about 50 percent would wean lambs and that the genetically superior half would wean lambs successfully. That half would be retained for replacements and the other half would be used for some other purpose. Afterwards every female that failed would be immediately removed from the selected group so that every female producing replacements would have weaned offspring at every opportunity.

Embryos from superior females from lambing at 1 and 2 years of age would be split and transferred to donor females (possibly those culled) to increase the number of offspring from superior ewes by a factor of 4 to 8 times. Super ovulation would not be used because it might interfere with selection for lambing rate. After the techniques were improved all replacements could probably come from embryo transfer or from younger mothers, thus reducing the generation length for females to three years or less. Identical siblings would be available for evaluation of meat quality and flavor. All females would be mated every 7.2 months to permit selection for removal of seasonality of reproduction. This will not detract from the ability for once a year lambing. Synchronization of ovulation, AI, and synchronization of parturition would be used in all matings not only for practical management, but also to obtain positive natural selection for success of these technologies.

Selection for success of AI from frozen ram lamb semen seems essential to improve the success of AI and to make possible the sale of semen rather than rams at some future time. The number of small flocks of sheep will probably increase greatly in the future as well as the number of large highly intensive-production flocks. Genetic improvement in these flocks and the management of breeding will both be enhanced by the use of AI. Success with frozen semen is quite variable and no doubt a part of this variability is genetic. Ram lambs will reach puberty early enough to permit semen to be frozen and then used in test matings with early pregnancy tests on the mated females. Thus, ram lambs could be evaluated for success of AI with frozen semen so that the superior males could be used as replacement males at about 7 months of age.

It is important not only that the best available adapted breeds be selected within each ecosystem area, but also that continued natural selection be applied within each ecosystem area. This will insure that productivity and adaptability will continue to be improved simultaneously. In the past, lack of adaptability was often not recognized or was covered up by hybrid vigor from crossbreeding. Improvement of adaptability is very slow. Therefore it is important to make some progress every year. Selective breeding within ecosystem areas is the best way to improve adaptability.

Sheep production requires special skills. Therefore, sheep enterprises should start small and grow from within. Mistakes are almost inevitable for the new sheep farmer. By starting small, the farmer tends to make mistakes when the consequences are limited. No one should undertake sheep farming just to make money. Rather it should be done for a satisfying way of life and to produce commodities essential for human life and well-being. Possibly in a human generation, many now part-time sheep farmers could make a living by raising small ruminants without taxpayer support.

Biotechnology research could increase biological efficiency of sheep and goat meat production from one to three percent per year in the next two decades and eventually to about five percent per year. These are generally far greater increases in efficiency than for other farm products and could also bring an increase in numbers possibly as great as ten percent per year. It would be expected that the production of meat from small ruminants might increase to about fifteen pounds per capita in two decades, while consumption of meat from other sources may decline to about half the current level. While sheep and goat meat would still be of minor importance, it would probably be the only low cost meat available, as it generally is in developing countries now.

SUMMARY

The farm problem of low income and high expenses has resulted from a transfer of income from farmers producing food to non-farmers working in the food chain and to consumers. This is resulting in a suppression of food production and a further reduction in the number of full time farmers unless government support returns an adequate standard of living to farmers. Meat and animal products are the most likely foods to be reduced, but meat and animal products are the ideal foods because they contain all of the nutrients essential to life.

Biotechnology research and development can increase the efficiency of meat production and can help

170

maintain the superabundance of food at relatively low
prices, which is the basis of our high standard of
living. This research will not give the spectacular,
exotic results usually presented by the news media but
rather will result from involving biotechnology in
improvement of ruminants by selection at a much more
rapid rate than has been possible in the past. Small
ruminants will likely enjoy the most rapid increase in
efficiency because they can produce high quality food
on forage alone, because they can utilize poor feed and
poor land better than cattle or other farm animals, and
because of their early age of puberty, short reproductive cycle, and high reproductive rate fits them for
effective use of biotechnology selection to increase
efficiency and net returns.

REFERENCES

Anderson, W. F. 1986. Genetic Engineering of Ani-
    mals. Basic Life Sciences 37:7-13.
Davidson, F. E. 1985. A president's view of bio
    technology. Journal of Animal Science 61:1281-1284.
Economic Research Service. 1986. Economic indicators
    of the farm sector. Washington D.C.: USDA-Economic
    Research Service.
_____. 1985. Costs of Production 1985.
    ECIFS, 5-1. Washington D.C.: USDA-Economic Research
    Service.
Hansen, M., L. Busch, J. Burkhardt, W. B. Lacy and L.
    R. Lacy. 1986. Plant Breeding and Biotechnology.
    Bioscience 36:29-39.
Louisville Courrier Journal. 1987. Scientists create
    new animals - and new fears among critics. The
    Courier-Journal, Louisville, Kentucky, 16 November.
Rexroad, C. E., Jr. 1986. History of genetic
    engineering of laboratory and farm animals. Basic
    Life Sciences 37:127-138.
U.S.News and World Report. 1985. Fresh from the labs,
    A farming revolution. U.S. News and World Report.
    25 November: 74-75.

Part III

International Perspectives

Robert E. Evenson

9. U.S. Agricultural Competitiveness: Evidence from Invention Data

The changing competitiveness of U.S. agriculture is now receiving much attention. Competitiveness as measured by costs of production depends on international financial factors, i.e., exchange rates, tariffs, and quotas, as well as on technology. This chapter will concentrate on the technology dimension of competitiveness.

This technology dimension encompasses both the capacity within a given country to produce improved technology for the farmers in the country and the inherent transferability of the technology. Transfer of technology means the degree to which cost reducing new technology in one region is also cost reducing in another region. For example, if a research program in the private or public sector develops one or more new higher yielding crop varieties that lower the cost of producing a bushel of that crop in Montana by 10 percent, that technology is fully transferable to Canada if it also lowers costs in Canada by 10 percent. It is partially transferable if it lowers costs by less than 10 percent and not transferable if it does not lower costs.

The producers in any given country then have an interest in facilitating "transfer-in" or "spill-in" of technology from other regions or countries. They also have an interest in inhibiting "transfer-out" or "spill-out" of technology to producers perceived as competitors. In general, investment in research capacity in the region helps to produce location specific technology for the region and to facilitate spill-in. Competition between regions then may take the form of research investment.

This chapter first reviews the conceptual issue of impediments to agricultural technology transfer and then turns to evidence from crop variety and invention data to examine four types of technology transfer in recent years between the United States and its competitors.

TECHNOLOGY TRANSFER

Transfer of technology can be thought of as being impeded by "interactions" with three classes of factors or characteristics of an economy:

1. Interactions with natural resources, such as soil and climate characteristics.
2. Interactions with economic conditions, as reflected in prices and scarcities.
3. Interactions with public infrastructure, particularly policy interventions.

The first class of factors is characterized by agronomists as genotype environment interaction, and a large number of studies have measured the change in the biological performance of plants and animals under different soil and climate characteristics. Virtually all plant material is subject to some degree to these interactions. The high yielding semi-dwarf wheats produced by CIMMYT in Mexico were not transferable to Brazil, for example, because of the high aluminum content of Brazilian soils. Plant breeders and crop improvement research programs are continuously designing plant types to be "tolerant" of particular soil and climate conditions. In Brazil, wheats are genetically selected to be tolerant of high aluminum levels. Elsewhere salt tolerance, drought tolerance, and insect and disease tolerance are being pursued by researchers. The resultant technology in a crop such as wheat is therefore highly location specific. In many crops there is little or no direct international transfer of certain types of technology. Indeed, even within a country, regional transfer may be limited.

Interactions with economic conditions are less apparent but are of great importance. Consider modern harvesting equipment such as the self-propelled combine. There may be no significant soil or climate interactions inhibiting its use in India, but the fact that hand harvesting labor can be hired for one or two dollars a day makes it uneconomic in India. Even significant technological improvements in the combine will not make it economic in low wage economies, and it will not be transferred to them because of this basic economic fact. Some machines, of course, are also affected by soil type and other natural environmental factors.

Price interventions and incomplete or poorly developed markets will also inhibit technology transfer. These tend to be less important, however, than the first two classes of interactions. Note here the distinction between transfer of technology and diffusion of technology. Diffusion of technology is the experimental process by which farmers learn which new technology is suited to their own particular producing environment. This process requires both skill and time and is facilitated by extension programs. It is an important process, but it is not of concern here. Direct transfer is really a characteristic of a new technology. It simply relates the economic value of that technology (e.g., the cost of producting a unit of output with the technology relative to the cost of using the next best alternative technology) in its country of origin to its value in other countries. (It can include the process of information exchange, but that is not the fundamental issue in transfer.)

Transfer, however, is not limited to direct transfer. This study will discuss four types or channels of technology transfer:

1. Direct transfer, as when a crop variety or an invention is used in a country other than its origin without further modification.
2. Adaptive transfer, as when a crop variety or an invention is modified by further breeding and adapted to the economic and climate conditions in another country.
3. Pre-technology science transfer, as when a scientific finding enables or induces inventions by applied researchers in another country.
4. Capacity transfer, as when the graduate science education obtained in the country of origin enables more effective and inventive-oriented research in another country.

Direct Transfer

This section examines data on invention. Inventions are discoveries of new technology that meet the following standards:

- They are new or novel
- They are useful
- They require an inventive step, i.e., they are "unobvious" to practitioners in the field.

Inventions may or may not be given patent protection. Invention can be classified as (1) mechanical, (2) electrical, (3) chemical, (4) biochemical or biotechnological, (5) managerial, or (6) policy. Traditional patent protection is offered primarily to the first three classes. (The U.S. patent classification system is organized in these three broad groups.) "Plant patents" providing patent protection to asexually reproduced plants were introduced in the 1930s but have not been used widely to stimulate private sector invention. In the 1960s in Europe, and in 1970 in the United States, a new legal instrument, the Plant Variety Protection Certificate, was instituted that provides patent-type protection to plant varieties. It has had a significant effect on the organization of inventive activity.

The Chakrabarty decision providing plant protection (or not ruling it out) to living organisms has extended traditional patent system protection to much of the modern biochemical or biotechnological invention. This too has had significant effects. Managerial technology generally does not enjoy legal protection except through copyright law.

Four data sets on invention deserve investigation here. The first data set is from the U.S. Department of Agriculture Plant Variety Protection Office. It reports crop varieties being registered for "patent-type" protection in the United States. Varieties of foreign origin are also identified by these data. The second data set is from the U.S. Patent Office. It covers patents granted in the United States in eight agricultural technology fields. This data set also identifies inventions of foreign origin being protected in the United States. The third data set is from DERWENT, Inc., an international patent information firm. The DERWENT data report patents in thirteen technology fields granted in seven countries by country of origin. (The technology fields in the U.S. data base do not exactly match those of the DERWENT data base because the fields must be defined in terms of U.S. patent classes in the U.S. data base and international patent classes in the DERWENT set.) The fourth data set is the international INPADOC data base. These data allow an examination of transfer over a longer period of time and for more countries than do the DERWENT data.

    <u>Crop Variety Data</u>. Prior to the 1970 Plant Variety Protection Act, only asexually reproduced cultivars were afforded patent protection in the United States. The 1930 Plant Patent Act provided limited protection and was not widely used. Since 1970 sexually reproduced cultivars have been given protection and the Act is now quite widely used. A number of European countries also provide this type of protection.

    Table 9.1 shows the source of a plant patents taken out in the United States between 1970 and 1986. One can readily see the sharply increasing role of the private sector in the major field corps. In general, the rate of patenting has increased over the period, particularly in wheat and soybeans. As yet, there has been little foreign interest in U.S. Plant Variety Protection certificates except in several of the grasses (Boyce and Evenson, 1975).[1]

    <u>U.S. Patent Office Data</u>. A second indicator of technology transfer is measured by the protection of foreign origin inventions in the United States. Eight agricultural technology fields using U.S. patent classes have been defined (Evenson, Putman and Pray 1985).[2]

    Table 9.2 shows that the total number of patents granted in these fields declined from the 1975-1979 period to the 1980-1984 period by 11 percent. The declines in earthworking equipment, post-harvest technology, and fertilizer were greater than this, indicating a possible diminishment of technological potential or a reduction in demand for these inventions. Inventions in harvesters, threshers, and biotechnology were higher in the second period than the first.

    The origin of the inventions varies considerably between fields. Individual independent inventors (i.e., those who have not assigned their inventions to a corporation) are important in the mechanical fields -- especially in animal husbandry -- but not in the

Table 9.1

Plant Variety Protection Certificates Issued in the U.S., 1971-1986

	Total Issued	% Public	% Foreign	Total Issued by Period				Public Certificates by Period			
				71-74	75-78	79-82	83-86	71-74	75-78	79-82	82-86
Field Crops											
Barley	37	.08	.19	0	12	2	23	0	2	0	1
Beans (Field)	24	.21	—	0	1	5	18	0	0	4	1
Corn (Field)	47	.02	—	0	1	6	40	0	0	1	0
Cotton	122	.13	—	22	31	38	31	1	6	6	3
Oats	17	.53	—	0	11	5	1	0	7	2	1
Rice	14	—	—	0	8	4	2	0	0	0	0
Safflower	6	—	—	0	1	4	1	0	0	0	0
Soybeans	393	.14	—	34	69	135	155	4	8	25	18
Wheat	145	.24	—	13	42	60	30	6	13	11	5
Total Field Crops	805	.154	.008	69	176	259	301	11	36	48	29
Grasses											
Alfalfa	40	.23	—	0	3	21	16	0	1	5	3
Bluegrass	26	.04	.31	0	6	9	11	0	1	1	0
Clover	8	.13	—	0	0	4	2	0	0	1	0
Fescue	50	.18	.40	0	5	16	29	0	3	2	4
Ryegrass	42	.05	.52	0	1	12	30	0	1	1	0
Other	7	.28	.57	0	1	4	2	0	1	1	0
Total Grasses	173	.141	.312	0	16	66	90	0	7	11	7
Vegetables											
Beans, Garden	121	—	.15	32	39	21	29	0	0	0	0
Beans, Lima	7	—	—	0	2	4	1	0	0	0	0
Cauliflower	16	—	.87	0	2	6	9	0	0	0	0
Lettuce	60	—	—	14	17	14	15	0	0	0	0
Mushroom	12	—	—	3	3	3	3	0	0	0	0
Onion	24	.17	—	1	3	8	12	0	0	0	4
Peas	175	—	—	20	48	46	61	0	0	0	0
Tomatoes	40	.10	—	0	9	12	19	0	0	0	4
Watermelon	19	.11	—	0	6	4	9	0	1	0	1
Other Vegetables	60	—	—	0	8	20	32	0	0	0	0
Total Vegetables	534	.019	.060	70	137	138	190	0	1	0	9
Flowers											
All Types	70	—	.142	17	24	16	13	0	0	0	0
All certificates	1592	.100	.064	196	353	479	594	11	55	59	4

Table 9.2
U.S. Patents Granted in Agricultural Technology Fields, 1975-1984

	Earth-working Equip.	Planters Diggers	Har-vesting Equip.	Thresh-ing Equip.	Animal Husbandry	Ferti-lizers	Biotech-nology	Post har-vest
Patents Granted								
1975-1979	554	128	339	83	807	1251	493	2866
1980-1984	451	120	418	96	786	1085	527	2340
Ratio, 1980-1984/1975-1979								
	.82	.94	1.23	1.16	.97	.87	1.06	.82
Proportion U.S. Corporation								
1975-1979	.38	.25	.50	.55	.24	.58	.40	.52
1980-1984	.36	.33	.48	.35	.24	.52	.42	.49
Proportion U.S. Government								
1975-1979	.00	.02	.00	.00	.01	.01	.03	.03
1980-1984	.01	.02	.00	.01	.01	.02	.02	.01
Proportion U.S. Individual								
1975-1979	.27	.36	.26	.12	.58	.03	.03	.12
1980-1984	.30	.28	.24	.18	.51	.02	.04	.13
Proportion Foreign Origin								
1975-1979	.35	.35	.24	.32	.17	.38	.54	.32
1980-1984	.34	.38	.28	.46	.24	.44	.52	.27

three chemical fields. The U.S. government has only minor importance in agricultural invention. Further, its imimportance in agricultural invention in the world system has declined. The central feature of these data, however, is the increase in the share of foreign inventions. In 1975-1979, 32.7 percent of U.S. patents in these fields were of foreign origin. (Of these, 3.65 percent were owned by U.S. firms engaged in U.S. research overseas.) By 1980-1984, 37.0 percent of these inventions were of foreign origin (4.8 percent owned by U.S. firms). Every field except earthworking equipment showed an increase in foreign origin shares. The increase was largest in the fields with the lowest foreign shares in the first period.

The foreign share can be considered an index of direct transfer since firms are protecting products and process in the U.S. market. By this measure the biotechnology inventions have the highest transfer potential with fertilizers and threshers next. In contrast to some of the crop varietal technology, virtually all these mechanical, electrical, and chemical inventions have a moderate degree of transfer (and transfer potential (Judd, Boyce, and Evenson, 1983).[3]

   International Patent Data (DERWENT). The issue of direct transfer can be addressed in a more complete and satisfactory way with international patent data. The DERWENT data base permit the definition of thirteen technology fields related to agriculture (USDA, 1985). These included three traditional mechanical fields (earthworking; planting, harvesting, threshing; and animal husbandry), two traditional chemical fields (fertilizers and agricultural chemicals), and four modern biotechnology fields (mutations and genetic engineering, microorganisms and tissue culture, enzymes, and apparatus and equipment). For each field a "trade" matrix was constructed for the United States, the United Kingdom, France, West Germany, Japan, Canada, and Brazil. It was not possible to obtain these data for Australia or Argentina. The trade matrix shows the number of patents granted in the field for the 1978-1984 period in each country according to the origin of the invention. Appendix A reports these trade matrices for each field.

   Table 9.3 reports a summary of data from the trade matrices. It calculates three indexes of transfer for each field, an index of net trade for each country, and an index of U.S. trade with the country.

   The three transfer indexes are:

A. The ratio of total patents granted in all countries to origin patents.
B. The ratio of U.S. patents granted in all countries to U.S. origin patents.
C. The ratio of U.S. patents granted in the United States to U.S. origin patents.

   Index C is the percent of U.S. patents granted to these six foreign countries. It shows essentially the same ranking of technology fields as the transfer index reported in Table 9.2.

   Index B shows another facet of transfer and is a good measure of technology export potential. High transferability of technology in a field has two facets. The first is that the technology is of value abroad and will lower costs if used. The second is that it can be "sold" through licensing or production abroad. U.S. inventors will capture rents

Table 9.3
Technology Transfer Indexes Summarizing Patent Activity Among Seven Nations, 1978-1984

Technology Field	Earth-working Planting	Harvest-ing Machinery	Animal Husbandry	Ferti-lizer	Agricul-tural Chemistry	Mutation Genetic Engineering	Biotechnology Microorganism Tissue Culture	Enzyme	Apparatus
(1) Total Origin Patents	2393	1570	1388	1774	13397	565	894	1098	1412
(2) Total Granted Patents	3563	2190	1907	2696	23814	1200	1691	1779	2321
(3) Transfer Index (A)	1.49	1.39	1.37	1.52	1.78	2.12	1.89	1.62	1.64
U.S. Export (B)	1.52	1.42	1.33	1.73	1.97	4.04	2.33	2.18	2.42
U.S. Import (C)	1.28	1.17	1.24	1.38	1.44	1.34	1.59	1.57	1.38
Tax Index									
U.S.	.24	.25	.09	.36	.54	2.69	.74	.62	1.05
U.K.	-.26	-.18	-.11	-.27	.60	-1.42	-.97	-1.04	-1.38
France	-.38	-.27	-.04	-.07	.82	-1.61	-.73	-1.23	-.98
Germany	.34	.09	.22	.35	.86	-.232	-.381	.08	.23
Japan	2.22	3.23	-.07	.00	-.29	-.413	.09	.05	-.11
Canada	-1.64	-1.53	-1.56	-1.46	-6.84	-.15	-7.30	-6.54	-7.33
Brazil	-.31	-.48	-.66	-2.90	-19.10	-41.0	-6.00	-5.75	-5.28
U.S. Trade									
U.K.	.025	.016	.029	.046	.067	.513	.149	.077	.129
France	.015	.032	-.004	.041	.045	.252	.072	.051	.122
Germany	-.019	.026	-.016	-.03	-.074	.261	.052	.000	.035
Japan	-.019	-.01	.016	.061	.220	1.356	.201	.179	.248
Canada	-.187	.144	.074	.137	.126	.104	.195	.195	.241
Brazil	.051	.041	.021	.101	.159	.286	.072	.051	.059

through the sale of technology. These rents, however, will be less than the value of the technology, and U.S. producers of the commodities may well lose more from the technology transfer than U.S. inventors gain in selling technology.

Perhaps the best general index is index A. By this index these thirteen fields are ranked according to tranferability (from highest to lowest)

1. Biotechnology (genetic engineering)
2. Biotechnology (tissue culture)
3. Agricultural chemicals
4. Biotechnology (apparatus)
5. Post-Harvest technology - meat - dairy
6. Biotechnology (enzymes)
7. Post-Harvest technology - fruits
8. Fertilizer
9. Post-Harvest technology - grains
10. Earthworking equipment
11. Harvesting - threshing
12. Animal husbandry
13. Post-Harvest technology - food preservation

The ranking by indexes B and C is quite similar. It is clear that modern biotechnology inventions have the highest transferability. Fertilizer inventions have medium transferability, and mechanical inventions have relatively low degrees of transfer. Both geo-climate and economic impediments operate most heavily on these latter forms of invention.

Table 9.3 also reports two indexes of trade patterns. The first is a general technology trade index for each country. It is defined as patents granted abroad to national inventors, minus patents granted at home to foreign inventors, divided by patents granted at home to national inventors. Since there is an artificially created trade balance in the data, some countries must be net importers. The table shows that the United States -- a net exporter of technology in every field -- is still the dominant force in international technology trade.

Canada and Brazil are large net importers of all types of technology. Canada produces a significant number of inventions, but its proximity to the United States makes transfer easy between the two countries. Since the United States is so much larger, it is natural for Canada to be a large importer. Brazil produces fewer patents and represents the case of "downstream" trade between developed and semi-industrialized countries.

Japan is a large patenting country. In fact it obtains more patents than the United States in most fields. It is overall a net exporter, although it imports in several fields. Japan outproduces the two traditional invention economies, France and the United Kingdom, in all fields except agricultural chemicals. Japan is also an exporter of some biotechnology and will probably become a large exporter in the future.

West Germany is also an exporter of technology in all fields, except two biotechnology fields where it appears to lag. Its performance as an exporter is second only to the United States.

France and the United Kingdom export only in agricultural chemicals. This is due to their long experience in the chemical industries.

The second index of trade patterns is trade with the United States. It is the ratio of the number of patents obtained by U.S. inventors in the country, minus the number obtained by that country's inventors in the United States, divided by the number of U.S. patents granted to national inventors. It has been noted that the United States is a net exporter in every field. This second index shows that, with several minor exceptions, it is a net exporter to every country. The United States exports significant technology in every field to Brazil and Canada, and it is a net exporter to all countries in biotechnology and post-harvest technology. (The United States does import significant agricultural chemical and fertilizer technology from West Germany.)

International Patent Data (INPADOC). The data set from the International Patent Documentation Center (INPADOC) provides a longer time period from which to look at patenting and to investigate transferability between countries. For this study, patenting in three fields and eight countries is presented. The fields are horticultural and plant materials (excluding sexually reproduced plants), agricultural chemicals (herbicides, insecticides, and biocides), and bio-chemical products. The countries include the leading patenting countries: the United States, the United Kingdom, France, West Germany, Japan, Switzerland, Italy, and Brazil.[4]

Table 9.4 summarizes the data for horticultural patents. Each panel is organized by priority country, i.e., originating country. West Germany thus was the origin country for 1,817 horticultural patents of which 1,486 were granted in West Germany and 331 in the other countries in the study (64 of these were granted in Italy and only 9 in Brazil). This field of invention thus has a low degree of transferability. The ratio of all granted patents to origin patents (Index A, Table 9.3) is 8364/6675 = 1.25. Note that this is lower than for any of the fields in Table 9.3. (The high domestic share in the table reflects this low degree of transferability also.)

The leading patenting countries are the United States, Japan, and West Germany. The United States shows a decline in its rate of patenting. Most other countries show little trend. Italy has a low rate of patenting in 1983 and 1984. In a field with a very low degree of transferability, invention tends to be related to the economic importance of the commodity affected in each country. Horticultural products (cut flowers, for example) tend to be quite important to a number of countries in Europe, and these transferability data have some policy relevance.

Agricultural chemical technology, on the other hand, has a relatively high degree of transferability. The transfer index for the 13,395 origin patents summarized in Table 9.5 is 31,862/13,395 = 2.38. This is a relatively high index (see Table 9.3). Because of high transferability, somewhat more specialization between countries is observed. The United States and West Germany are the leading origin countries. France, the United Kingdom, and Italy have lower shares of this market than of invention generally, while Switzerland has a high relative share (in fact on a par with their individual indexes of transferability [DOMSHARE]). The United States and Japan have higher domestic shares -- hence lower transferability than the European countries. Brazil does not produce transferable inventions. Switzerland and Italy lead this field in terms of transferability. (This is to some degree related to geographic proximity and country size.) Japan, Italy, and West Germany are increasing their invention in this field while the United States, Switzerland, and the United Kingdom show signs of slowing down in recent years.

Table 9.4
Horticultural Patents in Eight Countries, 1973-1985

Year	U.S.	W.Ger-many	Japan	Switzer-land	U.K.	France	Italy	Brazil	Total	Domestic share
Priority = West Germany										
1973	1	68	0	9	5	0	12	0	95	71.6
1974	1	99	0	9	4	2	5	0	120	82.5
1975	5	92	1	5	6	1	5	0	115	80.0
1976	6	130	0	9	6	4	9	1	165	78.8
1977	4	118	1	6	4	5	1	0	139	84.9
1978	3	142	0	14	10	11	2	0	182	78.0
1979	2	115	1	4	6	5	5	1	139	82.7
1980	3	124	0	9	6	3	3	0	148	83.8
1981	7	103	1	5	9	5	2	0	132	78.0
1982	2	114	0	7	7	3	0	0	133	85.7
1983	4	142	0	5	3	6	0	3	163	87.1
1984	1	108	1	2	0	9	0	3	124	87.1
1985	1	131	0	4	2	3	20	1	162	80.9
	40	1486	5	88	68	57	64	9	1817	
Priority = France										
1973	5	5	2	2	4	19	21	2	60	31.7
1974	4	13	0	4	4	26	20	1	72	36.1
1975	9	15	0	0	5	13	5	1	48	27.1
1976	3	11	0	3	3	55	1	3	79	69.6
1977	2	14	1	2	6	84	5	2	116	72.4
1978	2	7	0	2	3	62	5	2	83	74.7
1979	1	8	0	0	6	45	7	1	68	66.2
1980	4	5	0	1	1	51	2	1	65	78.5
1981	3	0	0	2	4	39	1	1	50	78.0
1982	3	2	1	1	2	30	1	1	41	73.2
1983	2	1	0	0	0	60	0	0	63	95.2
1984	1	4	1	1	1	74	0	0	82	90.2
1985	1	4	1	0	0	77	13	0	96	80.2
	40	89	6	18	39	635	81	15	923	
Priority = U.K.										
1973	1	7	0	1	26	3	1	0	39	66.7
1974	7	2	1	2	28	0	5	0	45	62.2
1975	1	5	1	3	33	1	3	0	47	70.2
1976	2	7	2	0	5	3	2	0	21	23.8
1977	3	4	0	0	20	2	0	0	29	69.0
1978	1	6	0	1	35	6	1	0	50	70.0
1979	2	4	0	0	32	1	2	0	41	78.0
1980	4	1	1	0	69	0	0	0	75	92.0
1981	3	1	0	0	79	2	0	1	86	91.9
1982	4	1	3	0	70	2	1	2	83	84.3
1983	3	1	1	0	27	3	1	0	36	75.0
1984	1	1	0	1	11	5	0	0	19	57.9
1985	2	1	0	0	29	2	7	0	41	70.7
	34	41	9	8	464	30	23	3	612	

186

Table 9.4 (continued)
Horticulture Inventions

Year	U.S.	W.Germany	Japan	Switzerland	U.K.	France	Italy	Brazil	Total	Domestic share
Priority = Switzerland										
1973	0	5	0	16	0	0	2	0	23	69.6
1974	1	6	0	10	0	1	0	0	18	55.6
1975	1	7	0	21	0	2	1	0	32	65.6
1976	1	7	1	13	2	0	2	1	27	48.1
1977	3	11	0	13	4	2	2	1	36	36.1
1978	3	8	0	25	1	4	0	2	43	58.1
1979	0	6	1	14	4	1	1	0	27	51.9
1980	1	3	0	9	1	0	0	0	14	64.3
1981	2	3	1	11	3	0	0	0	20	55.0
1982	1	3	0	6	1	0	1	0	12	50.0
1983	0	1	0	11	1	3	0	1	17	64.7
1984	1	0	0	12	0	3	0	0	16	75.0
1985	1	0	0	8	0	1	9	3	22	36.4
	15	60	3	169	17	17	18	8	307	
Priority = U.S.										
1973	149	10	3	0	4	2	10	5	183	81.4
1974	138	15	4	2	5	0	9	4	177	78.0
1975	133	12	4	0	7	1	9	2	168	79.2
1976	144	15	5	0	8	7	5	6	190	75.8
1977	153	15	2	0	10	14	3	2	199	76.9
1978	149	16	0	1	12	18	1	11	208	71.6
1979	105	15	4	0	11	8	5	5	153	68.6
1980	140	7	3	1	18	3	2	2	176	79.5
1981	128	4	1	3	9	2	1	0	148	86.5
1982	129	3	5	0	15	2	0	6	160	80.6
1983	77	4	1	0	3	5	1	5	96	80.2
1984	74	3	4	0	2	8	0	4	95	77.9
1985	86	5	4	0	6	7	8	10	126	68.3
	1605	124	40	7	110	77	54	62	2079	
Priority = Japan										
1973	9	1	0	1	2	0	0	0	13	0.0
1974	4	2	55	0	0	4	0	1	66	83.3
1975	1	4	157	0	0	1	1	1	165	95.2
1976	1	7	190	0	2	1	0	0	201	94.5
1977	5	8	129	0	1	3	1	0	147	87.8
1978	7	6	142	1	2	2	0	1	161	88.2
1979	6	2	144	1	4	1	0	2	160	90.0
1980	5	5	224	0	9	6	0	1	250	89.6
1981	5	0	101	0	4	3	0	0	113	89.4
1982	1	2	209	3	5	3	0	0	223	93.7
1983	1	3	166	1	4	7	0	0	182	91.2
1984	4	1	124	0	1	8	0	0	138	89.9
1985	5	5	141	0	2	4	4	1	162	87.0
	54	46	1782	7	36	43	6	7	1981	

Table 9.4 (continued)
Horticulture Inventions

Year	U.S.	W.Germany	Japan	Switzerland	U.K.	France	Italy	Brazil	Total	Domestic share
Priority = Italy										
1973	1	5	0	0	0	1	57	0	64	89.1
1974	2	7	0	4	1	0	78	0	92	84.8
1975	2	2	0	0	2	1	39	0	46	84.8
1976	0	2	1	0	0	3	25	0	31	80.6
1977	1	5	0	1	0	2	19	0	28	67.9
1978	0	3	0	1	1	6	16	1	28	57.1
1979	1	4	1	0	4	1	35	1	47	74.5
1980	3	3	1	1	2	0	9	0	19	47.4
1981	2	1	0	0	1	2	10	1	17	58.8
1982	3	2	0	0	0	2	9	3	19	47.4
1983	0	1	0	0	1	3	3	0	8	37.5
1984	0	1	1	1	0	5	1	0	9	11.1
1985	1	2	0	1	0	2	78	0	84	92.9
	16	38	4	9	12	28	379	6	492	
Priority = Brazil										
1973	0	0	0	0	0	0	0	0	0	
1974	0	0	0	0	0	0	0	6	6	100
1975	0	0	0	0	0	0	0	14	14	100
1976	0	0	0	0	0	0	0	10	10	100
1977	0	0	0	0	0	0	0	17	17	100
1978	0	0	0	0	0	0	0	14	14	100
1979	0	0	0	0	0	0	0	10	10	100
1980	0	0	0	0	0	0	0	9	9	100
1981	0	0	0	0	0	0	0	13	13	100
1982	0	0	0	0	0	0	0	11	11	100
1983	0	0	0	0	0	0	0	25	25	100
1984	0	0	0	0	0	0	0	10	10	100
1985	0	0	0	0	0	0	0	14	14	100
	0	0	0	0	0	0	0	153	153	

Table 9.5
Agricultural Chemical Invention Patents for Eight Countries, 1973-1985

Year	U.S.	W.Germany	Japan	Switzerland	U.K.	France	Italy	Brazil	Total	Domestic share
Priority = W. Germany										
1973	41	69	30	76	130	18	87	69	520	13.3
1974	43	120	14	56	93	27	64	112	529	22.7
1975	85	194	21	75	90	17	31	86	599	32.4
1976	89	131	9	82	109	67	9	80	576	22.7
1977	84	186	10	68	111	121	8	76	664	28.0
1978	97	210	19	65	118	120	5	100	734	28.6
1979	80	189	15	43	59	77	20	119	602	31.4
1980	88	243	39	29	92	60	50	133	734	33.1
1981	94	248	24	34	57	32	34	95	618	40.1
1982	85	253	68	79	28	46	8	79	646	39.2
1983	67	321	50	41	16	71	3	136	705	45.5
1984	69	293	32	25	5	22	0	102	548	53.5
1985	100	243	56	9	6	8	106	89	617	39.4
	1022	2700	387	682	914	686	425	1276	8092	
Priority = France										
1973	5	4	3	6	25	33	24	25	125	26.4
1974	13	13	5	7	26	47	14	15	140	33.6
1975	19	32	3	6	20	28	9	6	123	22.8
1976	14	23	3	11	30	57	3	16	157	36.3
1977	23	12	7	16	15	71	3	10	157	45.2
1978	19	12	5	31	18	48	1	20	154	31.2
1979	19	7	6	12	8	36	4	24	116	31.0
1980	14	5	5	4	12	43	15	32	130	33.1
1981	12	5	2	3	14	24	9	31	100	24.0
1982	13	14	3	4	14	27	10	25	110	24.5
1983	9	4	5	5	4	47	0	32	106	44.3
1984	12	6	5	5	3	46	0	47	124	37.1
1985	13	4	5	10	4	47	13	31	127	37.0
	185	141	57	120	193	554	105	314	1669	
Priority = U.K.										
1973	11	21	10	16	113	16	31	16	234	48.3
1974	7	30	4	20	73	13	24	24	195	37.4
1975	27	41	5	19	100	18	13	14	237	42.2
1976	31	38	5	24	92	32	8	23	253	36.4
1977	46	52	7	18	70	45	3	19	260	26.9
1978	49	49	4	21	79	41	1	32	276	28.6
1979	34	25	11	13	93	33	8	49	266	35.0
1980	49	15	16	10	130	42	24	35	321	40.5
1981	41	11	4	11	158	14	16	34	289	54.7
1982	25	10	8	16	109	20	9	15	212	51.4
1983	27	19	17	11	50	26	1	44	195	25.6
1984	40	10	8	18	37	20	2	33	168	22.0
1985	51	5	13	6	35	8	60	19	197	17.8
	438	326	112	203	1139	328	200	357	3103	

Table 9.5 (continued)
Chemical Inventions

Year	U.S.	W.Germany	Japan	Switzerland	U.K.	France	Italy	Brazil	Total	Domestic share
Priority = Switzerland										
1973	28	3	14	64	61	21	51	7	249	25.7
1974	30	45	7	88	43	27	39	19	298	29.5
1975	50	56	6	107	61	28	26	10	344	31.1
1976	56	40	2	113	62	38	7	22	340	33.2
1977	53	59	4	68	43	56	2	7	292	23.3
1978	48	64	9	102	39	58	0	11	331	30.8
1979	39	25	7	40	31	39	5	16	202	19.8
1980	39	21	11	33	24	54	15	10	207	15.9
1981	38	17	6	47	40	8	15	9	180	26.1
1982	50	26	12	41	22	11	5	10	177	23.2
1983	46	21	7	46	6	12	0	12	150	30.7
1984	47	5	7	35	8	3	0	19	124	28.2
1985	51	7	19	32	11	4	41	13	178	18.0
	575	389	111	816	451	359	206	165	3072	
Priority = U.S.										
1973	303	24	10	41	117	28	92	80	695	43.6
1974	367	76	11	35	88	39	57	94	767	47.8
1975	436	72	21	41	93	21	31	40	755	57.7
1976	356	70	21	52	104	49	13	72	737	48.3
1977	422	93	25	32	95	87	7	52	813	51.9
1978	452	77	29	38	80	101	4	55	836	54.1
1979	263	47	29	24	109	62	11	83	628	41.9
1980	384	34	38	37	115	69	41	76	794	48.4
1981	324	27	34	31	95	25	38	101	675	48.0
1982	320	41	34	47	88	40	17	66	653	49.0
1983	300	21	40	38	39	68	2	115	623	48.2
1984	307	19	47	37	38	40	3	86	577	53.2
1985	309	6	48	18	24	16	132	56	609	50.7
	4543	607	387	471	1085	645	448	976	9162	
Priority = Japan										
1973	24	7	26	16	48	14	38	31	204	12.7
1974	44	23	72	13	36	25	27	28	268	26.9
1975	37	41	257	12	23	15	9	18	412	62.4
1976	51	34	322	20	31	33	4	28	523	61.6
1977	57	22	283	20	34	34	2	11	463	61.1
1978	50	39	274	23	36	40	2	28	492	55.7
1979	30	43	203	14	47	28	4	42	411	49.4
1980	57	52	313	10	71	33	16	43	595	52.6
1981	62	26	195	11	38	30	11	27	400	48.8
1982	60	30	395	11	72	23	7	29	627	63.0
1983	44	18	283	22	32	53	1	44	497	56.9
1984	68	8	249	29	21	35	1	36	447	55.7
1985	76	11	370	15	11	25	43	42	593	62.4
	660	354	3242	216	500	388	165	407	5932	

190

Table 9.5 (continued)
Chemical Inventions

| | | W.Ger- | | Switzer- | | | | | | Domestic |
Year	U.S.	many	Japan	land	U.K.	France	Italy	Brazil	Total	share
Priority = Italy										
1973	0	0	1	3	2	1	26	1	34	76.5
1974	0	4	1	4	6	3	20	2	40	50.0
1975	2	3	0	0	3	2	9	2	21	42.9
1976	3	7	0	0	2	7	7	3	29	24.1
1977	7	7	1	0	2	6	4	2	29	13.8
1978	4	14	1	1	7	8	3	4	42	7.1
1979	3	16	0	1	7	4	7	6	44	15.9
1980	5	7	0	1	7	11	15	4	50	30.0
1981	6	9	0	1	8	5	14	11	54	25.9
1982	7	4	1	3	15	6	10	3	49	20.4
1983	9	8	0	6	9	27	1	11	71	1.4
1984	12	3	1	11	5	19	1	6	58	1.7
1985	9	1	0	5	3	12	76	2	108	70.4
	67	83	6	36	76	111	193	57	629	
Priority = Brazil										
1973	0	0	0	0	0	0	0	2	2	100
1974	0	0	0	0	0	0	0	14	14	100
1975	0	0	0	0	0	0	0	30	30	100
1976	0	0	0	0	0	0	0	14	14	100
1977	0	0	0	0	0	0	0	18	18	100
1978	0	0	0	0	0	0	0	20	20	100
1979	0	0	0	0	0	0	0	21	21	100
1980	0	0	0	0	0	0	0	16	16	100
1981	0	0	0	0	0	0	0	18	18	100
1982	0	0	0	0	0	0	0	15	15	100
1983	0	0	0	0	0	0	0	9	9	100
1984	0	0	0	0	0	0	0	16	16	100
1985	0	0	0	0	0	0	0	10	10	100
	0	0	0	0	0	0	0	203	203	

Table 9.6
Biotechnology Patents in Eight Countries, 1980-1985

Year	U.S.	W.Ger-many	Japan	Switzer-land	U.K.	France	Italy	Brazil	Total	Domestic share
Priority = West Germany
1980	13	20	12	7	13	0	1	3	69	29.0
1981	42	83	9	16	17	0	3	4	174	47.7
1982	26	116	36	16	6	0	3	6	209	55.5
1983	32	132	30	21	4	6	0	7	232	56.9
1984	34	131	13	8	2	5	1	9	203	64.5
1985	31	181	24	10	5	7	11	6	275	65.8
	178	663	124	78	47	18	19	35	1162	
Priority = France
1980	8	0	3	3	8	1	2	2	27	3.7
1981	16	7	3	0	14	19	2	0	61	31.1
1982	5	10	4	2	6	14	0	2	43	32.6
1983	16	3	10	7	3	44	0	7	90	48.9
1984	11	1	3	6	3	49	0	1	74	66.2
1985	6	2	3	5	5	93	11	4	129	72.1
	62	23	26	23	39	220	15	16	424	
Priority = U.K.
1980	11	6	17	1	49	0	4	4	92	53.3
1981	22	3	12	8	58	0	3	7	113	51.3
1982	34	13	16	10	52	0	8	5	138	37.7
1983	23	8	17	4	31	5	4	10	102	30.4
1984	17	3	13	6	42	8	0	5	94	44.7
1985	18	4	8	8	26	6	14	9	93	28.0
	125	37	83	37	258	19	33	40	632	
Priority = Switzerland
1980	5	1	3	3	6	0	2	1	21	14.3
1981	9	0	11	21	6	0	2	2	51	41.2
1982	5	1	5	13	2	0	0	0	26	50.0
1983	10	1	10	13	3	5	0	1	43	30.2
1984	10	1	2	10	1	3	0	1	28	35.7
1985	5	3	2	27	1	0	2	2	42	64.3
	44	7	33	87	19	8	6	7	211	

Table 9.6
Biotechnology Patents (continued)

Year	U.S.	W.Germany	Japan	Switzerland	U.K.	France	Italy	Brazil	Total	Domestic share
Priority = U.S.										
1980	148	14	32	8	76	0	4	7	289	51.2
1981	324	25	35	16	63	0	8	16	487	66.5
1982	301	48	86	17	56	2	12	16	538	55.9
1983	281	20	77	31	31	24	1	19	484	58.1
1984	255	36	27	16	29	25	2	30	420	60.7
1985	308	29	52	20	36	41	53	22	561	54.9
	1617	172	309	108	291	92	80	110	2779	
Priority = Japan										
1980	50	16	371	6	60	0	3	0	506	73.3
1981	73	47	472	16	57	1	18	3	687	68.7
1982	78	72	643	16	47	6	6	4	872	73.7
1983	78	40	428	21	51	51	1	7	677	63.2
1984	74	38	410	85	46	49	1	5	708	57.9
1985	101	39	531	10	34·	61	25	7	808	65.7
	454	252	2855	154	295	168	54	26	4258	
Priority = Italy										
1980	2	1	1	5	7	0	10	3	29	34.5
1981	10	6	1	5	7	0	6	2	37	16.2
1982	6	3	5	3	5	0	3	0	25	12.0
1983	6	4	1	6	6	6	2	0	31	6.5
1984	6	4	1	2	3	6	0	0	22	0.0
1985	3	2	1	5	0	6	31	0	48	64.6
	33	20	10	26	28	18	52	5	192	
Priority = Brazil										
1980	0	0	0	0	0	0	0	8	8	100.0
1981	0	0	0	0	0	0	0	22	22	100.0
1982	0	1	0	0	1	0	0	25	27	92.6
1983	0	0	0	0	0	0	0	16	16	100.0
1984	1	0	0	0	0	0	0	18	19	94.7
1985	2	0	0	0	1	1	0	9	13	69.2
	3	1	0	0	2	1	0	98	105	

The third technology field studied is the general field of biochemistry-biotechnology. This field of invention is so new that the International Patent Classes required to specify the field were not introduced until 1980. Thus there are fewer years to examine. This field has a transferability index of 9763/5848 = 1.67 and is thus intermediate in transferability between horticultural produces and agricultural chemicals (Table 9.6).

Japan dominates inventions in this field with the United States coming in second and West Germany a distant third. The field is rapidly growing and France shows the highest growth rate. The patterns of individual country transferability are similar to those observed for agricultural chemicals.

The INPADOC data thus tend to be consistent with the DERWENT data. Both sets show that transferability of technology can be measured and that it varies greatly by type of technology.[5]

Adaptive or Indirect Transfer

Evidence also exists for somewhat less direct forms of transfer of technology. Patent citation data tell something about the "intellectual parentage" of an invention.

U.S. patent examiners are required to cite references in establishing the validity (i.e., the novelty and inventive step) of patents granted in the United States. References may be prior U.S. or foreign patents or other references -- notably scientific papers. The principle is usually to cite the "next best" prior art to clarify the basis for the judgment of the novelty of the present invention. Citations are thus a kind of pedigree of the intellectual or technical parentage of the present invention. Citations may thus capture adaptive transfer.

Table 9.7 summarizes data for eight technology fields as defined from U.S. patent classes (see also Table 9.2) and reports the percent of foreign origin patents granted in the United States, citations per patent, citations of U.S. patents of foreign origin, and direct citations of foreign patents. The leading countries of origin of cites are also shown for each field. Also reported is the proportion of U.S. patent

Table 9.7

Table 9.7
Citations of U.S. Patents in Agricultural Technology Fields, 1975-1984

	Earth-working Equip.	Planters Diggers	Har-vesting Equip.	Thresh-ing Equip.	Animal Husbandry	Ferti-lizers	Biotech-nology	Post har-vest
Patents Granted								
1975-1979	554	128	339	83	807	1251	493	2866
1980-1984	451	120	418	96	786	1085	527	2340
Ratio, 1980-1984/1975-1979								
	.82	.94	1.23	1.16	.97	.87	1.06	.82
Percent Foreign Origin								
1975-1979	.35	.35	.24	.32	.17	.38	.54	.32
1980-1984	.34	.38	.28	.46	.24	.44	.52	.27
Citations/Patents								
1975-1979	9.52	17.01	5.71	13.19	15.67	7.53	12.64	7.08
1980-1984	11.62	20.45	7.46	13.54	16.83	9.43	13.11	9.68
Percent Foreign Cites(Indirect)								
1975-1979	.17	.10	.20	.08	.03	.20	.15	.13
1980-1984	.14	.09	.21	.11	.06	.20	.18	.17
Percent Foreign Cites(Direct)								
1975-1979	.12	.12	.05	.03	.01	.05	.02	.05
1980-1984	.15	.18	.10	.05	.03	.11	.03	.08
Percent Pre-1963 Cites								
1975-1979	.64	.72	.71	.79	.85	.52	.77	.50
1980-1984	.62	.70	.41	.71	.77	.46	.69	.44
Percent Science Cites								
1975-1979	.004	.011	.001	.000	.002	.103	.10	.06
1980-1984	.023	.011	.010	.003	.008	.100	.17	.09

cites that were pre-1963, which measures the dynamism of the field. Finally, a measure of the science orientation of the field is presented. This is the percent of cites to scientific papers.

Table 9.7 shows that citations per patent have risen in all fields over the two periods. For all fields, cites per patent are 10 percent higher in the later period. The proportion of pre-1963 cites has fallen over the two periods (from .645 to .588). This is primarily the result of the passage of time and not relevant to the analysis. This indicator does, however, suggest dynamism in a field. Short periods of citation as reflected in the post-harvest technology and ferti-lizer fields may indicate more rapidly growing fields.

The percent of science cites is relevant to this discussion (particularly in the section to follow) because it indicates whether the technology is science-linked. The data show significant links in fertilizers, post-harvest technology, and biotechnology. The biotechnology field is quite strongly tied to science in the later period.

These data show that the percentage of foreign cites went up in every field over the two periods. In the early period, 29.1 percent of all patents were granted to foreigners (foreign patents of U.S. ownership are not included), while 17.5 percent of all cites were to foreign patents. In the second period, 32.2 percent of all patents were granted to foreigners, while 23.6 percent of all cites were to foreigners. Thus the citation data are consistent with a growing foreign role in U.S. invention and with the recognition that foreign invention is a growing part of the intellectual structure of inventions.

Pre-Technology Science and Adaptive Transfer

The Agricultural Sciences. Patented inventions cover only part of the actual inventions relevant to U.S. agriculture. As noted earlier, many inventions are not patented. Even with strengthened plant patenting laws, most traditional biological technology -- plant and animal genetic improvements -- are not patentable. Data showing how the distributions of scientific papers by field have changed over time are informative regarding the changing capacity of applied agricultural research.

Table 9.8 reports scientific publication data from the Commonwealth Agricultural Bureau data base for twenty-four countries. These data are based on two 5-year periods (1973-1977 and 1978-1982), which are comparable, and changes over these periods can indicate changes in scientific output. Publications were classified into ten applied scientific fields as shown in the summary Table 9.9.

It should first be noted that the number of publications are not strictly comparable across fields. Publications in veterinary medicine, for example, may be less costly to obtain than publications in plant pathology. Most of these publications are traditional, although the veterinary medicine field probably contains many modern biotechnology science papers.

The U.S. shares and the ratios between the two periods are comparable across fields. The U.S. shares are generally high. The ratios show veterinary medicine, entomology/nematology, and soil science to be the most rapidly expanding fields, while plant breeding and animal nutrition have had reductions in scientific work.

The United States ranked number one in every field and produced a total of 289,061 publications in the CAB data set. The United Kingdom was second with 100,135 publications and it ranked second, third, or fourth in all fields. India followed with 89,750 and ranked second in the world in plant breeding, plant pathology, crop science, and soil science. This may seem a bit extraordinary, but reviews show India has done quite well in its agricultural performance. Projections made in the early 1970s that India would be importing large quantities (in excess of 50 million tons) of grain in 1985 have not materialized. This, incidentally, has had an impact on U.S. export markets. The USSR ranks next in general output. Canada and Australia, the two chief export competitors of the United States, both rank high in terms of scientific output. Italy ranks seventh in plant breeding and ninth in plant pathology.

The United States has increased its world role from 1973 to 1982 in animal breeding, soil science, crop science, plant pathology, plant breeding, and weed science. It has decreased its world role in animal nutrition, entomology, and veterinary science. The USSR, by contrast, lost ground in all fields except entomology and plant pathology. Argentina, Australia, Brazil, and Canada, the four major export competitors, expanded their roles in all fields except plant pathology and entomology. Developing countries (except for Egypt) have become more important in the agricultural science world. So have Japan and the exporting countries. Eastern Europe and the USSR have lost ground.

Pre-Agricultural Sciences: Changes in Capacity and Capacity Transfer

Table 9.11 summarizes changes in national agricultural research system investment from 1959 to 1980. These data are expressed in constant 1980 dollars. They show very significant shifts in research capacity.

Table 9.8
Total Publications, Percent of World Share, and Rank among 24 Countries
in 10 Scientific Fields, 1973-1982

Field	Argentina			Australia			Brazil		
	N	%	R	N	%	R	N	%	R
Animal Breeding	86	.217	24	1457	3.672	5	241	.607	19
Animal Nutrition	30	.098	24	1354	4.423	5	127	.415	20
Dairy Science	105	.242	22	992	2.284	9	269	.619	19
Veterinary Med.	736	.231	21	5258	3.807	6	1839	1.839	11
Plant Breeding	186	.381	23	809	1.658	11	863	1.769	10
Crop Science	162	.342	23	1610	3.395	7	1669	3.519	6
Weed Science	62	.318	23	515	2.642	7	319	1.636	10
Plant Pathology	77	.263	23	835	2.854	7	612	2.092	11
Entomol/Nematol	105	.228	24	1155	2.505	9	601	1.303	16
Soil Science	117	.383	24	1928	2.739	4	931	.958	10

Field	Canada			Czechoslovakia			East Germany		
	N	%	R	N	%	R	N	%	R
Animal Breeding	1122	2.828	9	1231	3.102	8	748	1.885	11
Animal Nutrition	1457	4.760	4	659	2.153	9	519	1.695	12
Dairy Science	853	1.964	13	1209	2.783	8	605	1.393	17
Veterinary Med.	4090	3.740	9	1810	1.309	12	1773	1.141	13
Plant Breeding	1384	2.837	6	795	1.630	12	640	1.312	14
Crop Science	1423	3.001	9	1142	2.408	10	563	1.187	15
Weed Science	845	4.335	4	238	1.221	12	226	1.159	13
Plant Pathology	971	3.318	6	362	1.237	14	411	1.405	13
Entomol/Nematol	1307	2.835	8	862	1.870	11	438	.950	19
Soil Science	1894	2.131	5	663	.943	14	578	.924	17

Field	Egypt			France			India		
	N	%	R	N	%	R	N	%	R
Animal Breeding	235	.592	20	1290	3.251	6	2209	5.567	3
Animal Nutrition	166	.542	19	899	2.937	6	1469	4.799	3
Dairy Science	436	1.004	18	1357	3.124	6	1607	3.699	5
Veterinary Med.	1158	1.209	18	5001	2.320	7	6138	7.749	3
Plant Breeding	230	.471	21	775	1.589	3	15499	11.272	2
Crop Science	605	1.276	14	999	2.106	11	15727	12.076	2
Weed Science	113	.580	21	435	2.232	9	912	4.679	3
Plant Pathology	277	.947	17	797	2.724	8	4229	14.453	2
Entomol/Nematol	633	1.373	14	1448	3.140	7	3079	6.678	3
Soil Science	612	.603	15	1175	2.605	9	3924	3.197	2

Table 9.8
Total Publications (continued)

Field	Israel N	Israel %	Israel R	Italy N	Italy %	Italy R	Japan N	Japan %	Japan R
Animal Breeding	279	.703	18	643	1.620	13	1234	3.110	7
Animal Nutrition	220	.719	17	422	1.379	13	826	2.698	7
Dairy Science	169	.389	20	874	2.012	12	1261	2.903	7
Veterinary Med.	747	.956	20	1712	1.173	15	4385	2.583	8
Plant Breeding	256	.525	19	1002	2.054	7	1451	2.974	5
Crop Science	399	.841	18	756	1.594	13	1824	3.846	4
Weed Science	178	.913	18	190	.975	16	442	2.267	8
Plant Pathology	262	.895	19	777	2.655	9	1352	4.620	4
Entomol/Nematol	255	.553	20	900	1.952	10	1471	3.190	6
Soil Science	484	.389	18	594	.892	16	1308	2.284	8

Field	Mexico N	Mexico %	Mexico R	Netherlands N	Netherlands %	Netherlands R	New Zealand N	New Zealand %	New Zealand R
Animal Breeding	156	.393	21	657	1.656	12	534	1.346	14
Animal Nutrition	223	.728	16	228	.745	15	296	.967	14
Dairy Science	121	.279	21	899	2.070	11	683	1.572	14
Veterinary Med.	530	.419	22	2104	1.677	10	1361	1.639	17
Plant Breeding	166	.340	24	506	1.037	15	223	.457	22
Crop Science	150	.316	24	439	.926	17	322	.679	20
Weed Science	44	.226	24	258	1.324	11	225	1.154	14
Plant Pathology	57	.195	24	468	1.599	12	330	1.128	15
Entomol/Nematol	136	.295	22	813	1.763	12	631	1.369	15
Soil Science	212	.276	21	849	1.096	11	830	.709	12

Field	Philippines N	Philippines %	Philippines R	Poland N	Poland %	Poland R	Sweden N	Sweden %	Sweden R
Animal Breeding	100	.252	23	773	1.948	10	471	1.187	16
Animal Nutrition	38	.124	23	644	2.104	10	195	.637	18
Dairy Science	42	.097	23	907	2.088	10	638	1.469	15
Veterinary Med.	192	.399	24	1758	1.438	14	1006	.529	19
Plant Breeding	286	.586	18	889	1.822	9	294	.603	17
Crop Science	442	.932	16	855	1.803	12	240	.506	21
Weed Science	157	.805	19	179	.918	17	211	1.082	15
Plant Pathology	129	.441	22	329	1.124	16	146	.499	21
Entomol/Nematol	167	.362	21	765	1.659	13	497	1.078	18
Soil Science	202	.100	22	728	.916	13	268	.524	19

Table 9.8
Total Publications (continued)

Field	Switzerland N	%	R	Taiwan N	%	R	U.K. N	%	R
Animal Breeding	312	.786	17	131	.330	22	2910	7.333	2
Animal Nutrition	80	.261	21	55	.180	22	2745	8.967	2
Dairy Science	632	1.455	16	37	.085	24	3051	7.023	3
Veterinary Med.	1372	.527	16	260	.300	23	12271	6.278	2
Plant Breeding	254	.521	20	349	.715	16	1837	3.765	4
Crop Science	229	.483	22	371	.782	19	3381	7.129	3
Weed Science	122	.626	20	68	.349	22	1724	8.844	2
Plant Pathology	188	.642	20	276	.943	18	1992	6.808	3
Entomol/Nematol	505	1.095	17	129	.280	23	3094	6.710	2
Soil Science	267	.715	20	152	.135	23	3179	6.392	3

Field	U.S. N	%	R	U.S.S.R. N	%	R	West Germany N	%	R
Animal Breeding	8579	21.620	1	531	1.338	15	1505	3.793	4
Animal Nutrition	6855	22.393	1	576	1.882	11	807	2.636	8
Dairy Science	7093	16.328	1	3180	7.320	2	2101	4.837	4
Veterinary Med.	29628	24.348	1	5547	3.412	5	5625	3.671	4
Plant Breeding	8702	17.837	1	3292	6.748	3	991	2.031	8
Crop Science	8960	18.893	1	1768	3.728	5	1462	3.033	8
Weed Science	6311	32.376	1	546	2.801	6	747	3.832	5
Plant Pathology	4943	16.893	1	1216	4.156	5	773	2.642	10
Entomol/Nematol	8946	19.402	1	2677	5.806	4	1564	3.392	5
Soil Science	10304	15.434	1	1728	2.890	7	1859	2.930	6

In general the developing countries have expanded their research capacity at an impressive rate. Research spending increased by a multiple of 5.8 in developing countries in Latin America, 6.9 in Asia, and 3.6 in Africa. Scientist man-year (SMY) multiples were 6.0 in Latin America, 4.1 in Asia, and 4.2 in Africa. This is in contrast to spending and SMY multiples for public sector agricultural research in the United States of 1.9 and 1.4, respectively. The major export competitors (Canada, Australia, Argentina, and Brazil) had spending multiples of 2.4, 4.0, 2.1 and 1.4, respectively.

Table 9.10 shows how research and extension "spending intensities," i.e., spending as a percent of the domestic value of agricultural product (G.D.P.) has changed from 1959 to 1980. These data show that in 1959 the low-income and middle-income developing coun-

tries were approximately twice as spending intensive
for extension as for research. The reverse was true
for the industrialized countries. The rapid growth in
spending intensities for research from 1959 to 1980,
combined with little or no growth in extension intensities in the 1970s, produced roughly equal spending
intensities for research and extension in most developing countries.

Table 9.11 provides comparable data for "manpower
intensities" (i.e., ratios of manpower to G.D.P.).
Developing countries fare better by this measure because spending per SMY is lower. The difference between the low-income and industrialized countries is
much reduced.

For extension, the picture is quite different. By
1959 low-income developing countries had attained very
high extension manpower intensities: 5.9 to 7 times
greater than those attained in industrialized countries. By 1980, with a slight decline in these intensities for industrialized countries, the difference was
even greater. Middle-income and semi-industrialized
countries also increased their extension intensities.

Table 9.9
Summary of Total Publications, Percent of World Share, and Rank
among 24 Countries in 10 Scientific Fields, 1973-1982

Scientific Field	Total Publications 1978-1982	U.S. Share	Total Publications 1973-1977	U.S. Share	Ratio 78-82/ 73-77
Animal Breeding	39680	.216	30435	.182	1.31
Animal Nutrition	30616	.240	39164	.255	.78
Dairy Science	43440	.163	36882	.163	1.18
Weed Science	19492	.328	14361	.303	1.09
Plant Breeding	48786	.178	50204	.161	.97
Plant Pathology	29260	.168	28030	.137	1.04
Entomology/Nematology	46113	.194	33126	.233	1.39
Crop Science	47424	.189	41722	.160	1.14
Veterinary Medicine	191965	.154	121319	.189	1.59
Soil Science	50658	.203	36096	.167	1.40

Table 9.10

Agricultural Research Expenditures and Manpower, 1959-1980

Region/Subregion	Expenditures (000 Constant 1980 US$)			Manpower (Scientist Man-Years)		
	1959	1970	1980	1959	1970	1980
Western Europe	274,984	918,634	1,489,588	6,251	12,547	19,540
Northern Europe	94,178	230,135	409,527	1,818	4,409	8,027
Central Europe	141,054	563,334	871,233	2,888	5,721	8,827
Southern Europe	39,212	125,165	208,828	1,545	2,471	2,686
Eastern Europe & USSR	568,284	1,282,212	1,492,783	17,701	43,709	51,614
Eastern Europe	195,896	436,094	553,400	5,701	16,009	20,220
USSR	372,388	846,118	939,383	12,000	27,700	31,394
North America & Oceania	760,466	1,485,042	1,722,390	8,449	11,688	13,607
North America	668,889	1,221,006	1,335,584	6,690	8,575	10,305
Oceania	91,577	264,037	386,806	1,759	3,113	3,302
Latin America	79,556	216,018	462,631	1,425	4,880	8,534
Temperate South America	31,088	57,119	80,247	364	1,022	1,527
Tropical South America	34,792	128,958	269,443	570	2,698	4,840
Caribbean & Central America	13,676	29,941	112,941	491	1,160	2,167
Africa	119,149	251,572	424,757	1,919	3,849	8,088
North Africa	20,789	49,703	62,037	590	1,122	2,340
West Africa	44,333	91,899	205,737	412	952	2,466
East Africa	12,740	49,218	75,156	221	684	1,632
Southern Africa	41,287	60,752	81,827	696	1,091	1,650
Asia	261,114	1,205,116	1,797,894	11,418	31,837	46,656
West Asia	24,427	70,676	125,465	457	1,606	2,329
South Asia	32,024	72,573	190,931	1,433	2,569	5,691
Southeast Asia	9,028	37,405	103,249	441	1,692	4,102
East Asia	141,469	521,971	734,694	7,837	13,720	17,262
China	54,116	502,491	643,555	1,250	12,250	17,272
World Total	2,063,553	5,358,595	7,390,043	47,163	108,510	148,039

Sources: Boyce and Evenson 1975; and Judd, Boyce, and Evenson 1983.

Table 9.11
Research and Extension Expenditures as a Percent of the Value of
Agricultural Product, 1959-1980

Subregion/Group	Public Sector Agricultural Research Expenditures			Public Sector Agricultural Extension Expenditures		
	1959	1970	1980	1959	1970	1980
Subregion						
Northern Europe	.55	1.05	1.60	.65	.85	.84
Central Europe	.39	1.20	1.54	.29	.42	.45
Southern Europe	.24	.61	.74	.11	.35	.28
Eastern Europe	.50	.81	.78	.32	.36	.40
USSR	.43	.73	.70	.28	.32	.35
Oceania	.99	2.24	2.83	.42	.76	.98
North America	.84	1.27	1.09	.42	.53	.56
Temperate South America	.39	.64	.70	.07	.50	.43
Tropical South America	.25	.67	.98	.34	.71	1.19
Caribbean & Central America	.15	.22	.63	.09	.18	.33
North Africa	.31	.62	.59	1.27	2.21	1.71
West Africa	.37	.61	1.19	.58	1.24	1.28
East Africa	.19	.53	.81	.67	.88	1.16
Southern Africa	1.13	1.10	1.23	1.64	.67	.46
West Asia	.18	.37	.47	.25	.57	.51
South Asia	.12	.19	.43	.20	.23	.20
Southest Asia	.10	.28	.52	.24	.37	.36
East Asia	.69	2.01	2.44	.19	.67	.85
China	.09	.68	.56	n.a.	n.a.	n.a.
Country Group						
Low-Income Developing	.15	.27	.50	.30	.43	.44
Middle-Income Developing	.29	.57	.81	.60	1.01	.92
Semi-Industrialized	.29	.54	.73	.29	.51	.59
Industrialized	.68	1.37	1.50	.38	.57	.62
Planned	.33	.73	.66	–	–	–
Planned - excluding China	.45	.75	.73	.29	.33	.36

Sources: Boyce and Evenson 1975; Judd, Boyce, and Evenson 1983; and USDA

Table 9.12
Research and Extension Workers to the Value of Agricultural Product, 1959-80

Subregion/Group	Ratio of SMY's per 10 Million Dollars Agricultural Product			Extension Workers per 10 Million Dollars Agricultural Product		
	1959	1970	1980	1959	1970	1980
Subregion						
Northern Europe	1.05	2.01	3.14	2.76	2.56	2.61
Central Europe	.80	1.21	1.56	2.19	2.77	2.73
Southern Europe	.93	1.17	.96	2.00	2.76	2.69
Eastern Europe	1.44	2.97	2.84	2.36	2.88	3.13
USSR	1.38	2.37	2.34	2.26	2.33	2.50
Oceania	1.91	2.64	2.43	2.26	2.17	2.11
North America	.84	.89	.84	1.44	1.31	1.08
Temperate South America	.46	1.15	1.32	.26	1.19	1.26
Tropical South America	.41	1.41	1.77	1.71	3.95	6.46
Caribbean & Central America	.53	.86	1.20	.82	1.53	3.12
North Africa	.91	1.44	4.24	18.83	28.45	22.23
West Africa	.33	.61	1.42	7.61	14.01	18.08
East Africa	.32	.77	1.76	16.28	22.41	26.64
Southern Africa	1.90	1.96	2.47	8.73	5.94	5.62
West Asia	.33	.84	.88	4.39	7.25	6.54
South Asia	.50	.65	1.29	20.83	19.51	19.53
Southest Asia	.47	1.28	2.07	9.81	13.07	19.72
East Asia	3.80	5.29	5.72	6.57	7.05	6.13
China	.22	1.66	1.49	n.a.	n.a.	n.a.
Country Group*						
Low-Income Developing	.43	.67	1.40	18.14	18.61	20.43
Middle-Income Developing	.69	1.31	2.40	8.89	14.68	15.98
Semi-Industrialized	.70	1.21	1.36	2.80	4.95	5.21
Industrialized	1.24	1.71	1.85	2.37	2.31	2.12
Planned	1.02	2.27	2.13	–	–	–
Planned - excluding China	1.40	2.54	2.50	2.29	2.49	2.63

Sources: Boyce and Evenson 1975; Judd, Boyce, and Evenson 1983.

Table 9.13
Research as a Percent of the Value of Product, by Commodity,
Average, 25 Countries, 1972-1979

Commodity	Africa	Asia	Latin America	All Countries	Spending by International Centers	Ratio IARC Spending to Total
Wheat	1.30	.32	1.04	.51	.02	.04
Rice	1.05	.21	.41	.25	.02	.07
Maize	.44	.21	.18	.23	.03	.11
Cotton	.23	.17	.23	.21	-	-
Sugar	1.06	.13	.48	.27	-	-
Soybeans	23.59*	2.33	.68	1.06	-	-
Cassava	.09	.06	.19	.11	.02	.15
Field Beans	1.65	.08	.60	.32	.04	.11
Citrus	.88	.51	.57	.52	-	-
Cocoa	2.75	14.17*	1.57	1.69	-	-
Potatoes	.21	.19	.43	.29	.08	.21
Sweet Potatoes	.06	.08	.19	.07	-	-
Vegetables	1.56	.41	1.13	.73	-	-
Bananas	.27	.20	.64	.27	-	-
Coffee	3.12	1.25	.92	1.18	-	-
Groundnut	.57	.12	.60	.25	.005	.02
Coconut	.07	.03	.10	.04	-	-
Beef	1.82	.65	.67	1.36	.02	.02
Pork	2.56	.39	.60	1.25	.02	.02
Poultry	1.99	.32	1.12	1.64	-	-
Other Livestock	1.81	.89	.42	.71	-	-

*Ratios are high because production is very low.

Sources: Judd, Boyce, and Evenson 1983; and USDA, Indices of Agricultural Production.

It is clear from these data that the capacity for agricultural technology production has changed drastically since 1959. The tropical and sub-tropical regions of the world had very limited capacity for technology production prior to 1959. Most of their capacity was directed toward "colonial crops" (tea, coffee, cotton, cocoa). Significant research capacity now exists throughout much of the world.

Table 9.13 shows the research emphasis in terms of research intensities in developing countries during the 1970s by commodity and by region. The table also shows the importance of International Agricultural Research Center (IARC) spending in these commodities. These data show that research emphases in developing countries, including Brazil and Argentina, are lowest in commodities that are not heavily traded: cassava, sweet potatoes, and coconut. The Latin American countries are investing heavily in wheat, rice, soybean, and citrus research. Developing country importers in Asia and Africa also have non-trivial research programs in these commodities.

Interestingly, livestock commodities receive more attention than crop commodities in general. The extensive IARC system has significant programs in the key export crops, rice, wheat, and maize. The success of these programs in producing high-yield wheat and rice varieties is well known.

SUMMARY

This study has provided as much quantitative measurement and meaning as possible to the concept of "transfer" of technology. Four classes or types of transfer have been identified: direct, adaptive, pre-technology science, and capacity. The review of the technology produced in recent years by what might be termed "conventional" agricultural research programs showed:

1. Relatively low levels of direct transfer for most technologies. Very few crop varieties or farm management techniques are internationally transferred because of soil and climate inhibitions. Mechanical inventions tend to have higher transfer potential, but often they are inhibited by economic factors. Some chemicals and reproducing methods (e.g., broilers) have had moderate transferability.

2. Indirect adaptive transfer has been more important. This transfer does not take place unless the receiving region has a research capacity. Public sector research systems have developed effective means for exchange of breeding materials and other types of scientific communication that aid this transfer. The general growth of agricultural research

systems, particularly in the LDCs has greatly facilitated this form of transfer. The building of the International Agricultural Research Center (IARC) system has further accelerated this transfer process. Of course, these growing research systems have built a capacity for research that is largely independent of transfer from the United States.

3. Pre-technology science transfer has become an increasingly important form of transfer as research systems throughout the world have expanded capacity. The United States is the leader in the production of most forms of agricultural technology. Its position in the production of pre-technology agricultural science findings is more dominant and these findings are broadly transferred throughout much of the world.

4. Capacity transfer has been important as well. U.S. aid has facilitated the building of research capacity in the developing countries. U.S. graduate training institutions have been responsible for training a large proportion of the agricultural scientists in the developing world.

The net economic impact of the production or transfer of agricultural technology over the past thirty or so years has produced broad trends. During the 1950s and 1960s the dominant position of the U.S. agricultural research institutions, both public and private, produced rapid productivity gains in the United States relative to the developing countries and the Eastern Europe-USSR bloc. U.S. comparative advantage was growing. By the early 1970s, the United States had expanded exports and benefited greatly from the Russian grain pacts and the oil-shock-induced price changes in the world market.

For roughly a decade now the expanded research capacity of developing countries (as well as some of the export competitors of the United States), along with the growing indirect transfer of technology, reversed the trend in growing comparative advantage for U.S. agriculture. This erosion of real comparative advantage has been masked by the international financial aspects of the "loose fiscal-tight monetary" policy of the United States since 1959. This policy has been very costly to U.S. agriculture; in addition, the comparative advantage of the United States has been eroded.

This chapter also assessed emerging technologies. Emerging technologies were grouped into two broad classes: those based on conventional methods and those based on modern molecular biology and related scientific developments -- biotechnology. These latter technologies have been stimulated by institutional developments enabling private firms to capture more returns from their research (chiefly by strengthened patent and plant variety protection laws).

The next decade will most likely be dominated by technologies emerging from the conventional systems -- although plant breeding will probably be shifted out of public sector programs to the private sector. The general trends of the past decade will continue to unfold. Productivity growth will be fairly high throughout much of the world. The Eastern Europe-USSR bloc will continue to fall behind the rest of the world. Western Europe will not aggressively expand research, except in mechanical and chemical technology where it will expand its influence.

During the next decade it is likely that the biotechnology sector will produce a number of significant products. This development will be volatile -- many failures, bankruptcies, and mergers will occur -- but by the end of a decade important new technologies will emerge. The United States will be in the vanguard in most of these technologies, which will emerge primarily from the private sector; public sector research programs in the United States and especially in the rest of the world will be far behind in many fields.

The emerging technologies will have a higher degree of direct transfer than older technologies. Indirect adaptive transfer will take place through multinational firms, but probably not through public research systems because of their slow transition to modern methods.

208

NOTES

1.   Since 1970 crop varieties developed in the
public sector agricultural experiment stations in the
United States have been registered in the P.V.P.A.
system.
2.   See Evenson, Putnam and Pray for details of
defining these technology fields. They are based on
the U.S. patent classification system.
3.   Foreigners will wish to obtain patent protection in the United States if the invention is useful
in the United States. Even if the invention is embod-
ied in a product to be exported to the United States,
the inventor will wish to protect this product from
copying by a U.S. manufacturer.
4.   See Evenson, Putnam and Pray for further discussion of this classification which is based on International Patent Classes.
5. An appendix detailing the number of patents
granted 1978-1984 for six priority countries for non-
fertilizer agricultural chemicals and post-harvest
technologies is available from the author.

REFERENCES

Boyce, J. K., and Evenson, R. E. 1975 National and
    international agricultural research programs.   New
    York: The Agricultural Development Council.
Evenson, R. E., Putnam, J., and Pray, C. 1985.   The
    potential for transfer of U.S. agricultural tech-
    nology. Report prepared for the Office of Technology
    Assessment Project on the International Competitive-
    ness of American Agriculture.
Judd, M. A., Boyce, J. K., and Evenson, R. E. 1983.
    Investing in agricultural supply.Paper   No. 442,
    Economic Growth Center, Yale University, New Haven.
United States Department of Agriculture, International
    Economics Division, Economics, Statistics, and Coop-
    eratives Service. 1985. Indices of Agricultural
    Production.

10. Application of Biotechnology and Other New Technologies in Developing-Country Agriculture

The world is at a tantalizing point in the evolution of science and technology. More so than at any previous time in history, there are two overwhelming and contradictory visions of the future. One is the vision of ultimate destruction as an increasing number of nations gain possession of nuclear weapons. The other is the vision of unlimited potential through biotechnology to improve the human condition beyond anything we have ever imagined. It is a challenge for this and future generations to ensure the continued existence of planet earth and to improve the quality of that existence.

During the twentieth century advances in science and technology have been more rapid than in any previous century. Technical abilities have grown remarkably as advances in all fields of science have been creatively fused. For example, knowledge from physics and chemistry has led to sophisticated medical diagnostic equipment that applies magnetic fields to identify specific compounds inside the bodies of humans and animals. With computer technology, this information can be used to reconstruct a three-dimensional image with which problem areas can be identified. Also, our increasing ability to communicate and travel allows access to places that were once out of reach, while manned and unmanned space flights have produced large quantities of data about the universe.

Agriculture is benefiting from advances in science and technology, not only in industrialized nations, but in the Third World as well. International efforts have been made to apply the new technologies to the serious agricultural problems in the developing world. Twenty-five years ago critical food shortages stimulated agricultural research that produced the "Green Revolution" with its high yielding wheat and rice varieties and improved management practices. Now a new arsenal of skills, collectively known as biotechnology, has arisen from advances in many fields, including molecular biology, genetics, immunology, and cell biology. On a small scale, these skills are already being used to attack the extremely difficult agricultural production problems that are still present in the developing world.

To determine, from both the global and U.S. perspective, how biotechnology can affect developing country agriculture, a number of questions must be answered. Why is biotechnology needed? What is the best way to apply biotechnology? How is the Agency for International Development (U.S.A.I.D) incorporating biotechnology into its overall stategy?

AGRICULTURAL PROBLEMS IN DEVELOPING COUNTRIES

Current world attention is focused on the agricultural problems in Africa, particularly in the Sahel region south of the Sahara where agricultural productivity is declining absolutely while the population continues to rapidly increase. While many of the same agricultural conditions are found in other developing regions, in much of Africa they are more numerous and more severe. A discussion of the problems that limit agricultural production in Africa, therefore, conveys some sense of the number and variety of constraints experienced throughout the developing world.

To begin with, while Africa has some highly fertile soils, there are vast areas where the tropical soils present management problems that have not yet been addressed. The majority of Africa's tropical soils are acid and infertile. When agricultural areas are expanded to accommodate increasing populations, the natural elements often cause erosion, depleted fertility, and reduced water holding capacity of the soil.

Africa has an additional constraint to agricultural production: it is mostly a dry continent. It has much lower overall water availability per hectare than either Asia or Latin America. Furthermore, Africa's food production capacities also are severely constrained by animal diseases and pests. On a continent where mechanical farm equipment is rare, animals are a major source of power for food production. Yet, a number of diseases prevent the production of livestock in large areas of Africa. One disease alone, the dreaded trypanosomiasis, which is spread by the tsetse fly, prevents livestock production on one-quarter of the continent. In the affected areas, human labor is the only source of power to till the land and carry out other agricultural practices.

Finally, African farmers do not have and, in some cases, cannot afford the levels of fertilizer and irrigation that were important in Asia's Green Revolution. Because water, financial, institutional, and human resources are limited, the small amount of African farmland that is now irrigated is not likely to be greatly expanded in the near future. The approach taken now must move beyond the technologies of the green revolution. Tools and methods must be developed to define and overcome the complex constraints that limit agricultural production. This is where biotechnology will play an important role.

BIOTECHNOLOGY AND DEVELOPING COUNTRY AGRICULTURE

Biotechnology gives scientists the ability to attack problems at a level not approachable before -- the molecular level. While constraints on agricultural production are most critical in developing countries, hopes are high that biotechnology will have a major impact on agriculture both at home and abroad.

In the animal sciences, the production of useful technologies is just about at hand. In the plant sciences, basic understanding of how the plant genome works needs to be enhanced, and useful products are further down the road. However, the first genetically engineered pesticide-resistant plants are now appearing and the road may not be as long as initially expected.

DNA probes and monoclonal antibodies can play a significant role in diagnosis and control of plant and animal diseases. Tissue culture will allow researchers to rapidly identify desirable qualities and move them into the field. These new abilities and techniques help scientists to address some of the more elusive constraints confronting agricultural production in the developing world.

During the last quarter of a century, as the importance of appropriate technology to the success of development assistance has become more evident, research has played a growing role in U.S.A.I.D's overall strategy. The Agency has begun and will continue to incorporate biotechnological methods into its research and development program.

Many of the problems affecting agriculture in the developing world have been identified and the potential that biotechnology offers to address these problems is impressive. But a difficult question remains: What is the best way to bring these two together? How can biotechnology be applied to the problems of a developing country in a reasonable manner that will produce advances on a realistic time scale and, at the same time, will contribute to the overall development of self-sufficiency and self-reliance in the Third World.

U.S.A.I.D is fully aware that the application of biotechnology to problems in the developing world requires adaptation and reinforcement. The Agency's approach is to support potentially high pay-off research programs while providing education and training to developing country scientists. In this way, biotechnology is being used to develop some of the agricultural technologies that are so badly needed at the same time that developing country scientists are being encouraged to prepare for participation in this exciting new scientific arena.

U.S.A.I.D's programs, both past and current, reflect these goals and strive to maintain a balance between the development and use of appropriate technology and the support and expansion of institutional and human resources in the developing countries.

Nitrogen-Fixing Bacteria

In 1975, U.S.A.I.D and the University of Hawaii initiated a joint effort to help to accelerate the use of one of the earliest biotechnologies in the developing world -- the application of biological nitrogen fixation(NiFTAL project). Nitrogen fertilizers are a significant and often prohibitive expense to farmers in the developing world. Inoculating legume plants with Rhizobium bacteria so that they can biologically fix nitrogen can significantly reduce this cost. Through the NifTAL project, U.S.A.I.D. is helping to reduce the dependence of small farmers on purchased nitrogen fertilizers for the production of food.

Scientists at the University of Hawaii are collecting, characterizing, cataloging, and storing <u>Rhizobia</u> samples from around the world. They are also developing an antisera bank for identification of the strains of <u>Rhizobia</u>. This is essential for optimally matching the legume with the rhizobium strain. These collections and related literature are now available to scientists worldwide. An inoculant development program has been started to generate systems for improved inoculant delivery and to ensure dependable, effective inoculation in the field. An international network of field trials has been put in place to determine the conditions under which the yields of tropical legumes can be increased by inoculation with selected strains of <u>Rhizobia.</u>

Training is a significant part of the NiFTAL project. Courses have been held every year on topics such as "commercial scale legume inoculant production" and "management of nitrogen-fixing trees and crops." Extension courses are also offered. Over the years, 200 linkages between people and organizations have been established through legume/rhizobium technology training courses held in Hawaii, Africa, Asia, and Latin America. In addition, forty-five interns from twenty-five countries have spent from several weeks to six months acquiring skills while working in collaboration with project scientists at NiFTAL.

The NiFTAL project has taken an appropriate technology and has improved and expanded its usefulness in the developing world. Now, as the project keeps pace with new scientific advances, similar efforts are beginning to take place in molecular biology. Research programs are being developed to improve <u>Rhizobia</u> at the genetic level through recombinant DNA technology.

Tissue Culture

Advances in plant tissue culture have moved rapidly and the potential impact of this field on agriculture is just becoming evident. U.S.A.I.D supports a cooperative effort with Colorado State University to use the methods of plant tissue culture to accelerate the development of food crop varieties with higher yields under conditions of stress. The ultimate intention is that small-scale farmers will grow such cultivars on developing country marginal lands. The project focuses on improving the stress-resistance of the important cereal crops: rice, wheat, millet, corn, and sorghum; as well as cowpeas and common beans, which are important legumes.

In collaboration with IRRI (International Rice Research Institute/Philippines), scientists at Colorado State have shown that exposing plant cells to increasing levels of salt can be used as a selection method for salt tolerance. The salt-tolerant plant cells can then be regenerated into whole plants that pass their improved tolerance to succeeding generations. Using tissue culture methods, rice strains with improved ability to tolerate saline soils were generated at Colorado State. These improved varieties are presently being field tested at IRRI.

The U.S.A.I.D-CSU project also trains scientists from developing countries in tissue culture methodology and has organized an international network to ensure that scientists throughout the developing world are kept informed of the latest advances and newest techniques.

Vaccine Research

The development of animal vaccines through biotechnology is one of U.S.A.I.D's new interests and the Agency is beginning to build a program on this topic. The use of vaccinia virus as a vector for animal vaccines holds particular promise. Such vaccines, which are heat stable and easily administered through scarification, would be well suited for tropical conditions. Additionally, vaccinia offers the potential for developing one vaccine for multiple diseases since multiple genes can be inserted into the virus.

EMPHASIS ON RESEARCH

As the importance of appropriate technology for successful development has been realized, research has played an expanding role in U.S.A.I.D's overall strategy. The scientific focus of many projects and participation in the International Agricultural Research Centers reflects U.S.A.I.D.'s interest in and understanding of the need for research. Now, as biotechnology gains momentum, this new scientific technology is being incorporated into many of the Agency's research efforts.

Collaborative Research Support Programs (CRSPs)

To more effectively gain access to the scientific expertise available in U.S. agricultural universities, U.S.A.I.D has set up eight Collaborative Research Support Programs (CRSPs) that cooperate with national research agencies. A number of the CRSPs are using biotechnology methods to find solutions to the major agricultural problems in developing countries.

Small Ruminant CRSP. For example, outbreaks of contagious caprine pleuropneumonia (CCPP) result in high mortality that constrains goat production in many countries. In Kenya, joint scientific efforts of U.S.A.I.D's Small Ruminant CRSP, the Kenyan Ministry of Agriculture and Livestock Development (MALD), and a U.S. university have yielded phenomenal progress toward the control of CCPP through development of a vaccine and a practical diagnostic test. Both of these breakthroughs have been achieved through biotechnological research programs. As soon as current field studies are completed, the easy-to-use diagnostic test and heat-stable vaccine will be made available for use in other countries of Africa and Asia where CCPP is a major problem for goat producers.

Peanut CRSP. The Peanut CRSP is also employing biotechnology. A breeding sub-project of the CRSP is using tissue culture techniques in an embryo rescue program to increase the frequency of fertile hybrids from crosses between wild and cultivated peanut species. During the last few years, significant progress has been made in regenerating plants from the callus of both wild and cultivated species. This use of biotechnology will greatly increase both the speed and success-rate of attempts to incorporate into cultivated peanut varieties the genetic resistance to unacceptable environment and pest conditions found in wild peanut varieties.

In cooperation with ICRISAT (International Crops

Research Institute for the Semi-Arid Tropics/India),
the Peanut CRSP is also using the latest biotechnology
techniques to isolate, identify, and contain a potentially disastrous peanut stripe virus accidentally
introduced into the United States. A non-destructive
ELISA (enzyme-linked assay) technique was adapted for
use in determining the presence of this seed-borne
virus. The technique allows researchers to identify
and remove infected peanut seeds. Non-infected seeds
can later be grown to maturity. The virus, which is of
worldwide interest for its present and potential impact
on agricultural production, is being successfully contained through international biotechnological cooperation.

Program for Science and
Technology Cooperation (PSTC)

Begun in 1981, the Program for Science and Technology Cooperation (PSTC) managed by U.S.A.I.D's
Science Advisors Office seeks new research ideas with
potential to solve serious problems facing developing
countries. Plant Biotechnology and Biotechnology/
Immunology are two of six areas of investigation
supported by PSTC's modest grants (2-3 years; up to
$150,000). Relevant research by the National Academy
of Sciences (NAS) is also funded by PSTC under six
research networks, one of which focuses on biological
nitrogen fixation. A good number of developing country
researchers and a few U.S. scientists have received
PSTC grants to carry out research in these areas and
some interesting and useful results are beginning to
appear.

International Agricultural
Research Centers (IARCs)

Scattered throughout the developing world are the
thirteen international agricultural research centers
(IARCs) that make up the Consultative Group on International Agricultural Research (CGIAR). These centers, a
quarter of the support for which is provided by
U.S.A.I.D, have very specific mandates. Biotechnology
now constitutes a small but important part of their
overall research programs.

Research at many of the centers involves exploitation of germ plasm for plant breeding and recent advances in plant biotechnology provide valuable new tools for these efforts. New techniques to manipulate plant and animal genetic material are being incorporated into many of these research programs. While much of the IARC research still uses traditional scientific methods, the techniques of biotechnology are being included at many levels.

The most intensive use of the new scientific techniques is taking place at the International Laboratory for Research in Animal Diseases (ILRAD), which is located in Kenya. ILRAD is using biotechnology techniques extensively in seeking ways to control two of Africa's most devastating animal diseases: sleeping sickness (trypanosomiasis) and east coast fever (theileriosis).

Trypanosomiasis restricts animal agriculture to areas that are free from infestations of the tsetse fly, thus severely limiting animal production in large areas of Africa. To gain access to desirable genetic traits from other African regions, embryo transplant techniques are being used to overcome legal constraints on cattle movement between countries. In trypanotolerance studies, ILRAD transferred embryos from disease-tolerant West African N'dama cattle into East African Boran cows. Three to seven weeks after experimental infection with trypanosoma congolense, 75 percent of the control group of Zebu calves required treatment to prevent mortalities while none of the N'dama calves required treatment.

ILRAD is also conducting research to produce vaccines against both sleeping sickness and east coast fever and has made significant progress on the latter. The new methods of monoclonal antibodies and recombinant DNA techniques are and will continue to play a central role in these efforts. Specific diagnostic tests are being developed to identify parasite species for both diseases. The methods of DNA cloning and sequencing are being used to understand the trypanosome's ability to vary its surface coat antigens and avoid the host's immune system.

Among the crop-oriented centers, several have concentrated their biotechnology staff and facilities in tissue culture laboratories. Such facilities may provide essential support and depth to the center's own research, as well as facilitating cooperation with other research institutions. Since vegetative material must be "cleaned up" before being sent to cooperating scientists, centers working on vegetatively propagated crops need to have tissue culture laboratories in order to exchange germplasm. CIAT (International Center of Tropical Agriculture/Colombia) and IITA (International Institute of Tropical Agriculture/Nigeria) are using thermal treatment or "thermotherapy" to develop pathogen-free cultures of cassava and sweet potato. This facilitates the worldwide germplasm exchange of these crops, which could once only be exchanged with difficulty because of quarantine problems.

At CIP (International Potato Center/Peru), in vitro culture and recombinant DNA virus testing have been used to effectively test for control of a number of viruses. These techniques have been adapted for use in regional and national programs and have opened a whole new range of conditions under which production of quality seed tubers is possible. Greatly increased distribution of disease-free clones has also been made possible through the use of these techniques. CIP is also exploring the use of protoplast fusion to transfer mitochondrial genes for male sterility from donor to recipient plants. It is hoped that this will eventually lead to an improved system of production of true-seed hybrid potatoes.

CIMMYT (International Maize and Wheat Improvement Center/Mexico) and IRRI both have made significant progress in wide cross research, which is seen primarily as a source of genes for resistance to disease and stress tolerance. IRRI has used embryo rescue to make interspecific rice crosses that were impossible in the past. At CIMMYT, wide crosses have been a part of the breeding programs in both maize and small grains for a number of years. Working with the University of Illinois, CIMMYT has developed a "transforming DNA technique," which allows transfer of small amounts of alien DNA. Further development of the technique may eventually lead to rapid and precise transfer of genetic information from distant species.

Other techniques are being evaluated for physical transfer of DNA. Agrobacterium will infect cassava, potato, and sweet potato, and therefore could be used for gene transfer through plasmid integration in the genome. Microinjection of DNA into egg cells and microspores may hold promise in some cereal species and is currently being looked at by IRRI. Such precise methods hold the advantage of introducing just a few desired genes, without an entire wild species genome.

Biotechnology techniques will remain important at the CGIAR centers. The extent of their use will continue to be determined by their relevance to a center's mission and the accessibility of center scientists and cooperating researchers for national research programs. In some cases, it will be prudent to rely on the expertise and capabilities found in laboratories in developed countries. However, when in-house biotechnology research capability will significantly enhance and complement a particular center's programs and goals, necessary resources should be made available.

ANOTHER NEW TECHNOLOGY

International Benchmark Sites Network
for Agrotechnology Transfer (IBSNAT)

While biotechnology is the most prominent of the new technologies being employed in development assistance research programs, another relatively new endeavor deserves mention. Through systems analysis, computer simulation and expert systems, the International Benchmark Sites Network for Agrotechnology Transfer (IBSNAT) is developing methods to help decision makers (including farmers) in developing countries to make better choices among alternatives.

Using agricultural simulation models, researchers can quickly and inexpensively predict the problems and potentials of crop productivity with particular combinations of soil type, environmental constraints, and management resources. By using this technology, countries can reduce the amount of site-specific research and avoid disastrous crop choices. IBSNAT is an international effort that employs state-of-the-art technical knowledge and information technology to help agriculturalists "fit" their crops to the conditions in which they farm.

CONCLUSION

The vast potential of biotechnology and other recent technologies are being explored and applied to solve developing country agricultural problems. But the scientific methods that have served researchers so well in the past are certainly not being abandoned. Rather, the older and newer research techniques are being blended in ways that it is hoped will produce better paths to successful agriculture, particularly for the small, rural farmers of the developing world.

U.S.A.I.D. is convinced that countries must have the scientific capacity to pursue new research efforts if they are to overcome natural constraints to development. It is unlikely that every Third World country will be able to support or carry out basic biotechnological research in the near future. However, a scientific infrastructure is not built overnight. It is a slow process, and one that must be sustained in order for developing countries to achieve the level of competency required to initiate the development and adaptation of technologies specifically appropriate to their needs.

In U.S.A.I.D, as in the larger scientific community, there is cautious excitement. Only a small segment of this new field's potential is evident. Because there are many complexities and unknowns, research will require long-term commitment of financial and human resources. Biotechnology applications that have obvious financial potential will certainly stimulate private-sector involvement in highly developed countries. Yet in terms of measurable impact on the human condition, the benefits of this emerging field can probably be greatest in the developing world.

11. From Green Revolution to Gene Revolution: Common Concerns About Agricultural Biotechnology in the First and Third Worlds

When laboratories first began to experiment with recombinant DNA techniques, a public outcry arose concerning the possibility that products of these laboratories might escape into the environment and leave surrounding communities vulnerable to "engineered" laboratory organisms. In fact, one of the first "biotechnologies," bovine growth hormone (BST -- bovine somatatropin),is about to spread through this country's dairy communities -- carried not by the winds, but rather by the marketplace. Two major ex ante impact assessments of bGH, one by Cornell University agricultural economists and another by the Office of Technology Assessment (OTA), have attempted to predict the effects of this new biotechnology on the dairy industry. Both studies, while employing substantially different primary data collection procedures, predict that bGH will lead to a major dairy industry "shakeout." [1] Both are based on a conceptual and policy model of technology adoption and diffusion that many experts have criticized for more than a decade.

The "classical" diffusion model argues that large farmers tend to be the most innovative and thus will adopt new technologies most successfully. Accordingly, social scientists working from the classical diffusion model view the less successful adoption of new technologies by smaller farmers, and ultimately their disappearance from agriculture, as being natural or inevitable. This model, however, has been called into question on several grounds: scientific validity, usefulness for policy purposes, and normative biases. By looking at criticisms of classical diffusion --

particularly in light of evidence from Green Revolution studies where this perspective has received the greatest scrutiny -- we suggest that the predicted "shakeout" associated with BST diffusion and use is far from inevitable. The basis for this assertion is a new diffusion paradigm, emerging both abroad and in North America, that strives for scale neutrality by redefining the role of institutional actors in the diffusion process.

THE EFFECT OF BST ON MILK PRODUCTION

Growth hormone is a protein manufactured in an animal's pituitary gland. It regulates how the body uses nutrients to manufacture protein, either for muscle or for milk. Increased levels of bovine growth hormone in a cow's blood stream signal to her lactational system that there are enough nutrients present to produce more milk. However, the hormone does not change the amount of feed a cow needs to maintain her own body. Forty percent of all nutrients in a cow's regular feed intake go to body maintenance. With BST injections, a cow eats more. This extra feed, however, goes largely to the production of more milk, without having to contribute more to the cow's maintenance requirements. Growth hormone therefore increases milk production by increasing a cow's efficiency in using nutrients to make milk. This means that each farmer gets more milk from the same number of cows (Bauman et al. 1985).

Without BST injections, a cow's lactation peaks 6-8 weeks after calving. Lactation then declines approximately 6 percent per month from this peak. The injection of BST during the lactation peak increased milk yields in test cows by 10-15 percent. However, in the late stage of lactation, increased milk yields averaged 30-40 percent above normal. In other words, BST "straightens out" the lactation curve so that the cow consumes feed and produces milk near peak levels all through the lactation cycle. In test cows, increased yield over the entire cycle averaged 25 percent (Bauman et al. 1985). Current estimates of actual average increases on farms are between 10 and 15 percent (Bauman et al. 1985). The increase on any individual farm will depend on the quality of the herd,

management strategies, and the adoption of other feeding management technologies, such as computerized feeding systems (Kalter 1985a, p. 130).

TECHNOLOGY DIFFUSION-ADOPTION:
THE CLASSICAL MODEL

Classical diffusion theory, the major method that
has heretofore been used to predict the socioeconomic
impacts of BST, employs a mathematical model based on
research that shows that the cumulative percentage of
farmers who have adopted a new technology follows a
"sigmoid curve" over time.   Classical diffusionists
divide the curve into three categories of users of a
technology: "innovators," "middle adopters," and "laggards." In this model, the innovative farmer is better
educated, has more resources to invest, and has a more
"modern" attitude about the world (Rogers and Shoemaker
1971; Rogers et al. 1988). Accordingly, being an
innovator has several prerequisites. These include
control of substantial financial resources to absorb
the possible loss due to an unprofitable innovation and
the ability to understand and apply complex technical
knowledge (Rogers and Shoemaker 1971).
The farmers adopting a new technology earliest
then reap the early windfall benefits ("innovator
rents").   The innovative farmer is thus rewarded for
adopting early by the short-term profits received
before the product, now increasing in supply, drops in
price (Cochrane 1979).
In this lower price situation, all farmers are
forced to adopt the new technology in order to remain
competitive. However, later adopters do not gain any
additional income; in fact, their net income may be
lower than before the technology was deployed. The
resulting marginalization of "laggard" adopters is, in
the classical view, a consequence of personal make-up
and outlook such as reticence to take risks and fear of
change:

Alienation from a too-fast moving world is apparent in much of the laggard's outlook. While
most individuals in a social system are looking to
the road of change ahead, the laggard has his
attention fixed on a rear-view mirror (Rogers and

Shoemaker 1971).

Other factors (poverty, lack of resources, lack of contacts with input merchants, or lack of use of institutionalized credit, and smaller farming plots) in the profile of late adopters, which might account for failure resources are likewise ascribed to psychological traits. Individual psychology among losing farmers thus becomes the mechanism by which questions of institutionalized inequity and impaired access are ignored.

In many ways, use of the classical diffusion model has achieved enormous successes. Until recently, it has been one of the most widely used social science paradigms. As Hayami and Ruttan (1985) state:

> The diffusion model of agricultural development has provided the major intellectual foundation for much of the research and extension effort in farm management and production economics since the emergence, in the last half of the nineteenth century, of agricultural economics as a separate subdiscipline linking the agricultural sciences and economics.

The same authors, however, go on to describe why this otherwise reasonable model has serious flaws:

> The limitations of the diffusion model as a foundation for the design of agricultural development policies became increasingly apparent, based explicitly or implicitly on the diffusion model, failed to generate either rapid modernization of traditional farms or rapid growth in agricultural output (Hayami and Ruttan 1985).

THE CRITIQUE OF THE CLASSICAL MODEL

In the last fifteen years, a number of researchers have criticized assumptions in the classical model.[2] The source of this change in attitude toward the classical model was due as much to lessons learned from extension experiences in promoting Green Revolution technologies abroad as to a reinterpretation of the role of smaller, poorer farmers in rural development.

It is no surprise that this change in extension science occurred roughly at the time when the poor were "discovered," and the initial worry about the "revolution of rising frustration" had given rise to the "Green Revolution" debate (modernization and incorporation as causes of poverty) and to McNamara's famous speech to the Board of Governors of the World Bank in 1973 (Roling et al. 1976).

Many analysts of the Green Revolution began to criticize the social psychological assumptions that selected "modern" farmers and ignored "laggards" (Roling et al. 1976). As Barker and Herdt observe (1984):

One cannot expect technological innovations introduced over a period of five years to modify a pattern of resource ownership derived from hundreds of years of history.

Others cited the disappointing results of several Green Revolution projects using the two-step diffusion approach -- i.e., the targeting of more "progressive" farmers first in hopes that other farmers would follow the lead. As Saint and Coward (1977) suggest:

The Green Revolution called into question many of the tenets of the diffusion paradigm . . . The adoption process seemed limited more by technological and institutional factors than by the traditional barriers to communication, including illiteracy, fatalism, rural values, and lack of media exposure.

The controversies over the distribution of benefits from the Green Revolution have prompted a number of studies, the results of which argue against the assumptions of the classical diffusion model. These studies show first that small farmers are not necessarily "laggard" in relation to technological change [3] or, given adequate institutional support, they are often able to catch up quickly.[4] Second, the nature of technological diffusion depends as much on the technology to be diffused and the diffusing institutions as it does on attitudes held by adopting farmers (Burke 1979; Lipton and Longhurst 1985).

Third, small farmers can adopt new technology efficiently.[5] Fourth, nonadoption may have as much or more to do with the social and institutional setting farmers are working in as it has to do with individual innovativeness or management factors.[6] And fifth,the damage done where extension agents target large farmers outweigh the benefits of this cheaper, "efficient," and more gratifying extension policy.[7]

In sum, critics -- including "reformed" classical diffusion theorists such as Rogers -- fault the classical model for: (1) reductionism, which obscures alternative technological choices and scenarios; (2) ethnocentrism, which takes for granted that superior technologies originate in dedicated research centers, flow from center to periphery, and are improvements on indigenous technologies; and (3) the model's focus on individual motivations rather than institutional constraints.

THE NEW MODEL

In response to these critiques, development specialists have revised the classical model and called for the formulation of a new diffusion paradigm that places greater emphasis on farmer participation and equity. This new model recommends: (1) more emphasis on adaptation instead of adoption -- i.e., technological choice from a variety of alternatives suitable to different kinds of farmers (Saint and Coward 1977); (2) a farming systems, research-to-technology development orientation, predicated on a two-way information exchange between farmers and researchers (Roling et al. 1976; Rogers and Kincaid 1981); (3) a non-normative "supply side" approach that views all farmers as potential innovators and avoids the self-fulfilling prophecy that well-endowed farmers are more successful and less risk averse (Brown 1981); and (4) admitting the constraints of the political and institutional context in which smaller farmers work and tailoring these to the needs of farmers with diverse management skills and resource endowments (Gotsch 1972; Saint and Coward 1977; Roling et al. 1976). Many, though not all, demonstrations of the new model's effectiveness come from Third World settings. According to Ruthenberg (1985):

A great deal of empirical evidence suggests that small farms -- despite their small size and their initial low level of productivity -- have considerable capacity for change and improvement, that is, for agricultural development in the broadest sense. For the mobilization of this capacity, both technical change and institutional change are required.

THE BST STUDIES

Both the Cornell and OTA studies utilize the classical diffusion model in their reports on the effects of BST. In line with classical assumptions, the reports argue that large farmers will be "innovative" and adopt BST technologies earlier than small- to medium-sized farmers. The Cornell report states, "large herds are indicative of better managers, who can be expected to be more innovative and greater risk takers" (Kalter et al. 1985). The Cornell team conducted a survey of New York State dairy farmers and focused on when they planned to adopt BST. As Rogers has noted, however, there is a difference between measuring plans to innovate and the successful adoption of a technology. Moreover, both BST reports confuse the early willingness to take risks with ultimate survival. New diffusion studies in the Third World have shown that smaller farmers, while not necessarily adopting early, can adopt quickly and successfully, despite the windfall profits accruing to early adopters. [8]

More specifically, the quantitative computer models projecting adoption used in both studies make the assumption that the largest farms will use BST first. As a result, they show that the smaller commercial farms, many of which are currently viable, will disappear with the introduction of BST. These models suggest that the extent of the total farm disappearance depends primarily on the on-farm results of the hormone, management skills, and government price support policy in response to surging milk supplies. Predictions of dairy farm losses go as high as 50 percent over the next 10 years in the Northeast and Midwest, where dairy farms tend to be smaller (OTA 1986).

Both the OTA and the Cornell researchers acknowledge the possible consequences of widespread diffusion of the hormone -- increased unemployment, erratic land market shifts, family stress, and other social traumas (Kalter and McGrath 1985; Kalter 1985a). However, the social costs associated with retraining and relocating displaced farm operators, providing counseling and social services, and dealing with depressed property values are not incorporated into the cost-benefit analyses in the studies.

Like other classical studies in this area, diffusion is defined primarily in terms of individual behavior and attitudes, without significant attention to the role of institutions and social or political factors in the diffusion process. Both reports repeatedly warn that an "orderly exit" of resources from dairy production is necessary to avoid what the Cornell report calls the "rather drastic effects that could result if markets are allowed to clear" (Kalter et al. 1985).

Cornell's extension policy recommendations emphasize the need to retrain exiting dairy people, provide psychological counseling, and otherwise cushion the effects of failure to adopt BST (New York State Senate 1985). Yet an important empirical question remains to be answered: Do the major barriers to successful adoption of BST lie with the farmers themselves or with the institutional context in which they operate?

CLASSICAL DIFFUSION AS A SELF-FULFILLING MODEL

Certainly, to be forewarned of upheaval in the dairy industry is worthwhile -- indeed, social scientists in land-grant universities have given far too little attention to ex ante research. Yet many readers of the Cornell and OTA BST studies have tended to forget that these studies are forecasts, and forecasts are notorious for being inadequate, incomplete, politically biased, or simply wrong (Greenberger et al. 1977; Dutton and Kreaemer 1985). When OTA concludes that all the milk now sold by the United States could be produced by less than 5,000 well-managed dairy farms of 1,500 cows each (OTA 1986), one sees an extreme example of forecasting grounded in classical diffusion assumptions.

Though many argue that the classical assumptions accurately describe the real world of market places,

individual risk assessment, and personal motivations
for early versus late adoption, there is a danger that
these assumptions can become self-fulfilling.   In the
case of dairying, lack of critical thinking with regard
to these assumptions has led to the dairy herd buyout
programs cooperative extension policies that emphasize
cushioning the blows of an "orderly exit" rather than
exit prevention, and BST research in all major American
universities being funded by 4 or 5 agrochemical firms
with virtually no funds allocated to the social impacts
of the biotechnology.   Thus, currently prosperous dairy
operations will benefit most from the commercialization
of BST and other biotechnologies because public and
private institutions have cast them as the innovators
and early adopters of new agricultural technologies
(OTA 1986).

If the adoption of BST is not to be a self-fulfilling prophecy in terms of distributional effects,
important institutional changes will have to occur.
Though the neo-diffusion paradigm goes beyond institutional change, the absence of such change can undermine
all other tenets that reformers have set forth.   The
remainder of this chapter elaborates on three areas of
institutional reform that, if implemented, might dramatically influence the character of who wins and loses
as BST diffuses across the nation's dairy community.

INSTITUTIONAL INNOVATIONS

Extension Service

The world's best endowed system of agricultural
diffusion is the USDA's Federal Extension Service in
coordination with the 50 state agricultural extension
services (Rogers 1983).   Early in 1987 the National
Agricultural Research and Extension Users Advisory
Board submitted a report to congress and the President
urging a shift in philosophy regarding research, extension, and industry regarding farm technologies wherein
they recommended (1987):

> New technology should be available to all
> farmers, regardless of skill level or farm size,
> in order to increase their profitability and de-
> crease their dependency on price support pro-

grams.... The key to making the best use of available new technology will be to select for transfer only those innovations which will be of continuing economic benefit not only to the agricultural industry but also to consumers and taxpayers.

Though directed at the extension service in the United States, because of concern that biotechnology in agriculture may bypass marginal small farmers interest (Pimentel 1987), the essence of this recommendation echoes reform strategies advanced in a variety of Third World societies as well.

First among these is the insightful critique of extension in East Africa by Roling et al. (1976), which pinpoints the need among extension agents to overcome their prejudgment of which farmers are "progressive" and thereby deserving of extension attention. Also commenting on Green Revolution developments abroad, Wolf (1986) cautions the facilitators of agricultural policy against equating low income with low yields among peasants. This perspective mirrors the views of Schultz and others who testify to the rationality of peasants and small producers -- those who are too easily passed over by conventional extension philosophies and applications (Blair 1971). CIMMYT's manual, "Planning Technologies Appropriate to Farmers," directed principally at Third World extension organizations, has much for the United States and other First World extension services to emulate. Its 1980 preface states:

...a number of recent experiences have shown even the poorest farmers -- presumably the most tradition bound and usually those with least access to information, inputs and markets -- adopt certain technologies while rejecting others. Based on research on the diffusion of new cereal technologies in many countries, our own experiences and the reports of many others, we concluded that farmers do not adopt recommendations because they are not suitable for them.

The CIMMYT authors elaborate on the necessity of adopting technologies for diverse farmer needs and compile their manual to help extension agents accomplish this objective. Still others emphasize that diffusion tends to be inclusionary to the extent that farmer participation is stressed, both in the design of new technologies and in the design of actual field tests related to them (IDRC 1984; Ashby 1986).

Some may respond to the examples cited with skepticism precisely because they derive from Third World circumstance and may not apply to extension and other Land Grant conditions in the United States. This issue is addressed at length later in this analysis. Yet a domestic example warrants attention, one closely associated with the subject matter of Griliches' classic diffusion research, hybrid seed corn. Similar to what is projected for modern biotechnologies, the spread of hybrid corn varieties in the Depression years was swift. Land planted in hybrids rose from 500,000 acres in 1935 to over 24,000,000 acres in 1939, but not because of a classical diffusion bias in producing and disseminating the seed. Instead, the experiment stations and extension programs in several states trained "large numbers of farmers to participate in the highly technical job of producing hybrid seed corn" (Kloppenburg 1984). The University of Wisconsin, the centerpiece in this experiment, developed administrative procedures and special farm machinery to make this happen (Crabb 1947).

Private Sector

Reforming public extension services, though important, will strike some as too little too late. This is because much extension activity at home and abroad has been privatized, a trend likely to continue as biotechnology displaces older technologies developed by Land Grant institutions (Andrade 1984; Ruttan 1982). Such a trend does not preclude institutional reform, however, especially when defined as the equalization of risk and benefit across many classes of farmers. An early (and not entirely successful) example of private sector research and "extension" on behalf of smaller farmers was the cotton harvestor developed by the Rust brothers (Street 1957). At the other extreme there are the efforts of Control Data Corporation, a Fortune 500 company, to underwrite a demonstration project in Minnesota to show that farms of 150 acres can compete in today's farm marketplace (MacFadyen 1984).

It is in Third World settings where partnerships between agribusiness and other private sector actors and local producers of diverse description has made the most headway. This is true in the area of technology development and transfer (IRRI 1985), in production partnerships (Williams and Karenoffers 1986), and in what is referred to as an "industrial extension in agriculture" (Singer 1982). Such initiatives on the part of private sector actors are predicated on two forms of self interest.

In North America and in many developing nations, small farmers typically constitute high percentages of farm operators and require generous amounts of nonlabor inputs to increase productivity (Morris 1983). Second, under certain conditions, the "campesinista" (pro small farmer) perspective has gained private sector adherents by those who believe that farmers with varying resource endowments have rationally adapted to their environments by employing mixes of land, labor, capital, and indigenous knowledge that can not be improved upon by outsiders (IDRC 1984). Farming systems research itself is an institutional reform prompted by such perspectives in both public and private sectors (Harwood 1979; Shaner et al. 1982).

Other institutional reforms will be necessary in the private sector if the biases embedded in the classical diffusion model are to subside. Currently, for instance, the execution of environmental and social impact assessment for biotechnologies requiring approval by the Food and Drug Administration (FDA) is at the discretion of FDA. The likelihood of scientific risks being revealed through such assessments is low, however, since FDA cannot disclose proprietary information submitted by the private sector unless the companies publically disclose the information themselves (Schulman 1987). This means that the government will rule on the release of new technologies without the benefit of public debate informed by key proprietary research. Consequently, statements by BST researchers to the effect that BST can benefit smaller dairy operators (Kalter 1985b) require no empirical evidence subject to public scrutiny and are therefore unlikely to be incorporated in farm trials of the growth hormone.

Public Sector

Of the two reports discussed earlier in this chapter, the OTA analysis goes farthest in proposing institutional reform. "To assure a diverse, decentralized farm structure, where all sizes of farms have an opportunity to compete and survive," the OTA authors recommend that Congress draft various programs targeted to large, medium, and small farm operators, including multi-tiered price supports (Tangley 1986). Affirming the importance of institutional change in the public sector, the report's director comments:

> ...public policy is the most important driving force behind the transition in agriculture today...through policy, we can develop almost any kind of agricultural structure we want (Tangley 1986).

It is not in vain, therefore, that Senator Albert Gore, Jr. (D-Tenn.) recently declared that "biotechnology will be a hollow victory for science and for society if only the big boys survive to divide the spoils" (Gore 1987). Gore has recommended institutional innovations such as a biotechnology extension service that would "consciously steer progress toward the little guy" and a rural development bank that would offer low interest loans for small farmers to invest in biotechnology.

Like the private sector, the public sector may be disinclined to abandon preferential support for larger farms. With limits on arable land abroad increasingly evident, Edward Schuh of the World Bank reasons, Third World governments have increased motivation to substitute technology as a solution to their food and fiber production needs and trade imbalances. He adds that U.S. determination to remain internationally competitive in agriculture and to ease its balance of payments plight suggests a similar urge to reach for new technologies and old diffusion/adoption formulas at home. Compelling as these motivations may seem, research on small farms can be part of public sector strategies to raise yields and remain competitive (World Bank 1975; Berry 1984; Wolf 1986).

Tobacco production in the United States offers an example of government policies that elevated the needs of small farmers above those of "technological imperatives." This was accomplished through the Agricultural Adjustment Act of 1938, which through its "allotment system" preserved the tobacco belt of the South as perhaps the best remaining example of the Jeffersonian ideal of an independent yeomanry (Berardi 1984).

By 1970, however, labor intensive flue-cured tobacco production was irrevocably altered by the diffusion of mechanical harvesters developed jointly by the R.J. Reynolds Tobacco Company and the Agricultural Research Service faculty at North Carolina State University. Anticipating such mechanization, Congress passed P. L. 87-10 in the 1960s, which authorized intracounty lease-and-transfer of allotments, thus permitting the consolidation of quotas into larger management units established earlier. Research in 1980 on farmers who had exited tobacco production indicates the lease-and-transfer payments constituted "a large portion of the yearly household income for allotment holders, indicating that public policy helped small farmers share the benefits of tobacco mechanization in a variety of ways" (Berardi 1984). [9]

A second case of exemplary public policy aimed at equalizing the distributive effects of new technology arose in the Columbia River Basin where widespread irrigation became available after completion of the Grand Coulee Dam. In 1941, the Department of Interior (DOI) executed an _ex ante_ impact study to discover what farming strategies would be most successful when irrigation was adopted. In accomplishing this, DOI investigated assorted farm characteristics, including farm size, in eight federal irrigation projects in the Northwest. Farm size was shown not to be an important indicator of success. Thus, the government promoted irrigation services among all size farmers in the Columbia Basin (Bureau of Reclamation 1942). An equivalent strategy with regard to forthcoming biotechnologies would be _ex ante_ impact assessments involving field tests of new products on different sized farms to test the classical diffusion tenet that larger farmers will put such technologies to best use.

Numerous government policies can be proposed to offset the large-farmer bias of most diffusion efforts. The multi-tiered subsidy system proposed by OTA, despite its problems, is an excellent example. Sustained commitment of sufficient funds to federal research and development (including the social and environmental impact analyses of biotechnologies) in agriculture is

another example. Currently, less than 2 percent of
federal research and development spending goes to agriculture out of an annual $60 billion research budget,
despite the fact that agriculture remains the nation's
largest business (Young 1987). On this matter, Ruttan
(1986, p. 1) refutes those who argue that agricultural
research is frivolous in a time of surplus production:

> It is important for both producers and con-
> sumers that the agricultural research mission not
> be too narrowly defined... Research should also be
> directed to the design of more efficient institu-
> tions to protect both our production capacity and
> the income of farm people from the costs resulting
> from the integration of U.S. agriculture into
> world markets. Society should also insist that
> agricultural research be concerned with the ef-
> fects of agricultural technology on the health and
> safety of agricultural producers, with the nutri-
> tion and health of consumers, with the impact of
> agricultural practices on the esthetic qualities
> of natural and modified environments, and with the
> quality of life in rural communities.

AMERICAN EXCEPTIONALISM

Throughout this analysis, we have urged that lessons from abroad, especially lessons from the Green
Revolution, be applied to the diffusion of "gene revolution" technologies at home. Legitimate objections
may be raised to this message in light of Third World
ecological, cultural, and factor endowments, etc.,
which have little in common with the United States.
While differences such as these are of demonstrable
importance, there remain major commonalities in the
diffusion processes and their impacts on both
industrialized and industrializing agricultural systems.

Studies of classical diffusion in the United
States repeatedly show that mechanization has caused
massive technological unemployment [10], depopulation
of rural areas, and the decline of rural communities
and culture. And, as in other societies, the introduction of many new agricultural technologies has led to
greater concentration of agricultural production into

larger farm units.[11] In point of fact, exactly how
the effects of new technologies should differ between
less developed countries and developed countries is not
clear in the diffusionist literature, thus inviting the
type of analysis advocated here.

At best, for all its unique qualities, U.S agriculture is only a partial exception to the diffusion
lessons accumulating abroad. The important point is
that   lessons   can   be   learned   regarding   the
distributional effects of classical diffusion regardless of where they originate, given the commonalities
that invariably exist. To underscore this point, we
conclude our discussion by turning to Japan, the locus
of a self-styled "Green Revolution" with several implications for diffusion theory and practice in the United
States.

JAPANESE FARM TECHNOLOGY

In their work on agricultural development, Hayami
and Ruttan make three points of particular relevance to
the present study. First, "biotechnology" in agriculture is not new, despite recent advances in genetic
engineering, nor does it necessarily migrate from the
industrialized west to other nations (Hayami and Ruttan
1985). Japan, in the wake of the Meiji Revolution,
enthusiastically imported farm technologies from, and
modeled their experiment stations after, the United
States. By the end of the last century U.S. technologies were deemed inappropriate by the Japanese,
however, and thus were replaced with biologically-based
technologies suited for intensive, small-scale
agriculture. The results, even without economies of
scale and agglomeration common in the West, led to a
prodigious increase in productivity.[12]

Second, the Japanese case demonstrates that biological technology generally entails divisible inputs
such as improved seed and fertilizer and usually requires intensive on-the-spot supervisory management
decisions (Hayami and Ruttan 1985). Its effect, Hayami
and Ruttan argue, has been to raise the relative efficiency of small family farms and promote a unimodal
farm-size distribution. Many observers assume that the
structure of Japanese agricultural holdings were
grossly skewed prior to the onset of its post-World War

II land reform. The contrary is true. Japan's conscious choice of biological agricultural technology with a scale preference for middle-sized farmers minimized differences in farm size.

Third, the Japanese have not embraced the assumption, ingrained in classical diffusion theory, equating large farmers with "progressive" farmers. After discontinuing Western demonstration projects in machinery, livestock, and plants, Japan perfected a local extension service employing veteran farmers as itinerant instructors. The system was designed to diffuse the best seed varieties already in use and those that local farmers could combine. The seed was further refined and developed through a highly decentralized national experiment station system. Thus, Hayami and Ruttan (1985) call attention to the institutional innovations that accompanied and ultimately made a success of this veteran farmer initiative. They note that although experiment station-based research evolved in many·new biological directions as the benefits of local plant breeding and extension ran their course, this original model still influences research, development, and diffusion in Japanese agriculture.

Do these observations from Japan, like those selected from the more recent Green Revolution abroad, bear on the diffusion of BST in the United States?' We believe they do. These observations suggest that the self-fulfilling nature of classical diffusion tenets can be avoided. This requires a determined consideration of alternative technologies and an evaluation of the distributive effects on different classes of farmers at the same time as efficiency studies are undertaken. In addition, these observations suggest that biotechnologies in agriculture may be more amenable than mechanical technologies to widely dispersed benefits and that field tests should be structured to rigorously test this notion. These observations also suggest that the biases of the original diffusion model will not be rectified without emphasizing broad-gauge, on-going institutional reform (Binswanger and Ruttan 1978).

The promise of biotechnology for a full spectrum of agricultural producers is not limited to Japan. Lipton (1985) shows that modern plant varieties are especially favorable to smaller farmers, hired workers, and even poor consumers in developing countries.

Indeed, Norman Borlaug contends that the Rockefeller Foundation's original objective in supporting Green Revolution research was overridingly to help poor peasant farmers (Wright 1984).

In Japan extensive institutional scaffolding was arranged to see that the potential of biological innovations was realized by a cross-section of its farmers. Much of the Green Revolution's "pro-poor potential" was lost, however, through its insertion into social systems "favoring urban groups and big farmers who supply them" (Lipton 1985). Here again commonality with the American social system seems obvious, though one imagines the United States has more means at its disposal than most societies to alter such conditions.

A FINAL WORD

This analysis is, at base, about technological determinism. Such reductionism does not occur in a vacuum, but rather in the context of social philosophies that accord superiority to technologies that are new and relatively efficient from a labor-saving standpoint. Efficiency, thus defined, has its proper place in certain times and places, but is not a universally valid axiom in the United States or abroad. Nor should the labor displacing effects of previous technologies be used to rationalize further labor displacement by biotechnologies. BST proponents, for example, cite the 77 percent decline in U.S. dairy farms over the last two decades as justification for an innovation that will advance this trend (Tangley 1986). Were BST proponents to reverse this logic and pursue research showing how biotechnologies such as BST could halt such a trend by assisting small and medium dairy operators to remain competitive, then a philosophic pillar in America's on-going farm crisis will have been pulled.

NOTES

1. See R. J. Kalter et al. (1985) and Office of Technology Assessment (1986). The former used a sample survey technique to assess the profiles of farmers intending to adopt BST at particular points after com-

mercial introduction. The latter employed a "consensor" technique for making a comparable assessment of
the configuration of the "adoption curve" from the
judgments of experts in animal agriculture technologies.

2. For example, Roling et al. (1976); Saint and
Coward, (1977, pp. 733-737); Gotsch (1972); Rogers and
Kincaid(1981); Holden (1972); Brown (1981); Chambers
(1983); and Cernea et al.(1984).

3. See Prahladachar, (1983); Harriss (1977);
Ruttan, (1977); Barker (1983); M. Lipton (1985).

4. Burke (1979); de Walt (1978); Gladwin (1979);
Brown (1981); Roling et al. (1976); Gotsch (1972); and
Byerlee and de Polanco(1986).

5. Gotsch (1972); Perrin and Winkelmann (1976);
Griffin (1974); Prahladachar (1983); Williams and
Karenoffers (1986). Structural variables influencing
the adoption process include infrastructure, credit,
ethnicity, social connections, and the role of government (ranging from tariffs and price supports to land
reform and extension activities); for a "systems" approach, see Rogers and Kincaid (1981), who define a
system as "a set of interrelated parts coordinated to
accomplish a set of goals" (p. 46) and emphasize that
"person-blame rather than system-blame permeates most
definitions of social problems. Seldom are the
definers able to change the system, so they accept it.
Such acceptance encourages a focus on psychological
variables in social science research" (p. 40).

6. See McAllister (1981).

7. See Barker and Herdt (1984).

8. The Cornell researchers defined "large herd"
as 72 dairy cows, a size virtually indistinguishable
from those defined as mid-term adopters. More important, "large herd" is unquestionably a small herd by
all standards advanced in the OTA report for both the
United States and New York State. The Cornell research
cites Rogers' earlier studies but do not acknowledge
his more recent criticisms of research concentrating
exclusively on the predispositions of individual
adopters.

9. Other examples of special programs originating
in the New Deal period and intended to make technology
available to limited resource farmers appear in Larson
(1947).

10. For examples, see Horace Hamilton (1937); P.
S. Taylor(1938, pp. 595-607); R. I. McWilliams (1971);

A. Schmitz and C. Seckler(1970, pp.569-77).
    11.   See N. D. Fligstein (1978); A. Bertrand,
(1958); J. M. Brewster, (1958, pp. 1596-1608).
    12.   See R. J. Savill (1941); H. F. Breimyer and
W. Barr (1972); B. Gardner and R. Pope (1976, pp. 295-302).

REFERENCES

Andrade, E. R. 1984.   Notes on dissemination of   new
    technology.   In Brazilian agriculture and agricul-
    tural research,   ed. L. Yeganiantz, pp. 221-246,
    Brasilia, Department of Diffusion of Technology.
Ashby, J. 1986.   Methodology for the participation of
    small farmers in the design of on-farm trials.
    Agricultural Administration 22(March-April):28-39.
Barker, R. 1983.   Adoption, spread and production
    impact of modern rice varieties in Asia.   Los
    Banos, Philippines: IRRI
Barker, R., and Herdt, R. W. with B. Rose.   1984.
    The rice economy of Asia.   Washington, D. C.:
    Resources for the Future.
Bauman, D. E., Eppard, P. J., DeGeeter, M. J., and
    Lanza, G. M.   1985.   Responses of high producing
    dairy cows to long-term treatment with pituitary and
    recombinant somatotropin.   Journal of Dairy Science
    68:1352-1361.
Berardi, G. M.   1984.   Can tobacco farmers adjust to
    mechanization?   A look at allotment holders in two
    North Carolina counties.   In The social consequences
    and challenges of new agricultural technologies, eds.
    G. Berardi and C. Geisler, pp. 181-196, Boulder:
    Westview.
Berry, R. A.   1984.   Land reform and the adequacy of
    world food production.   In International dimensions
    of land reform, ed. J. D. Montgomery, pp. 63-82,
    Boulder: Westview.
Bertrand, A. 1958.   Agricultural technology and rural
    social change.   Chapter 26 in Rural sociology: an
    analysis of contemporary rural life,   ed. A. Ber-
    trand.   New York: McGraw-Hill.
Binswanger, H. P., and Ruttan, V. W. eds.   1978.
    Induced innovation -- technology, institutions and
    development.   Baltimore: Johns Hopkins University
    Press.

Blair, H. W. 1971. The green revolution and 'economic man': some lessons for community development. Pacific Affairs 44(3):44-59.

Breimyer, H. F., and Barr, W. 1972. Issues in concentration versus disperson. In Who will control U. S. agriculture?. Special Publication 27. Urbana-Champaign, IL: Univ. of Illinois, College of Agriculture, Coop. Extension Service.

Brewster, J. M. 1958. Technological advance and the future of the family farm. Journal of Farm Ecology 40(December):1596-1608.

Brown, L. A. 1981. Innovation diffusion: a new perspective. New York: McThuen Brown.

Bureau of Reclamation. 1942. Columbia Basin joint investigations. Washington, D. C. U. S. Department of the Interior.

Burke, R. V. 1979. Green revolution technologies and farm class in Mexico. Economic Development and Cultural Change 28(1):135-154.

Byerlee, D., and de Polanco, E. Hesse. 1986. Adoption of technological packages. American Journal of Agricultural Economics 68(3):519-527.

Cernea, M. M., Coulter, J. K., and Russel, J. F. A. (eds.). 1984. Research-extension-farmer. Washington, D. C.: The World Bank.

Chambers, R. 1983. Rural development: putting the last first. London: Longman.

CIMMYT. 1980. Planning technologies appropriate to farmers. Mexico D. F., Economics Program -- Centro Internationale de Mejoramiento de Maiz e Trigo.

Cochrane, W. W. 1979. The development of American agriculture: a historical analysis. Minneapolis: University of Minnesota Press.

Crabb, R. A. 1947. The hybrid corn-makers: prophets of plenty. New Brunswick: Rutgers U. Press.

Dasgupta, R. 1977. The new agrarian technology in India. New Delhi: MacMillan.

de Walt, B. 1978. Appropriate technology in rural Mexico. Technology and Culture 19(1):32-52.

DuPuis, E. M. 1985. No limits to growth. California Farmer (September 7):6-7, 17-18.

Dutton, W. H., and Kreaemer, K. L. 1985. Modeling as negotiating. Norwood, N.J.: Ablex.

Fligstein, N. D. 1978. Migration from counties of the south, 1900-1950. Ph.D. dissertation, Department of Sociology, University of Wisconsin, Madison, WI.

242

Gardner, B. D., and Pope, R. D. 1976. How is scale and structure determined in agriculture?" American Journal of Agricultural Economics 60(May):295-302.

Gladwin, C. H. 1979. Cognitive strategies and adoption decisions: A case study of nonadoption of an agronomic recommendation. Economic Development and Cultural Change 28(1):155-75.

Gore, Albert Jr. 1987. National Research Council News Report 37(2).

Gotsch, C. 1972. Technological change and the distribution of income in rural areas. American Journal of Agricultural Economics 54(2):341.

Greenberger, M., Crenson, M. A., and Crissey, B. L. 1977. Models in the policy process. New York: Russell Sage Foundation.

Griffin, K. 1974. The political economy of agrarian change. Cambridge, Mass.: Harvard University Press.

Hamilton, Horace. 1937. Population changes in Texas. (mimeo) College Station, Texas: Texas A & M College Agricultural Experiment Station, Division of Farm and Ranch Economics.

Harriss, J. 1977. In Green revolution, ed. B. Farmer. London: MacMillan 139-40.

Harwood, R. R. 1979. Small farm development -- understanding and improving farming systems in the humid tropics. Boulder: Westview.

Hayami, Y., and Ruttan, V. W. 1985. Agricultural development. Baltimore: Johns Hopkins University Press.

Holden, D. E. W. 1972. Some unrecognized assumptions in research on the diffusion of innovations and adoption of practices. Rural Sociology 37:463-469.

IDRC. 1984. Coming full circle: farmers participation in the development of technology. Ottowa.

Kalter, R. J. 1985a. The new biotech agriculture: Unforeseen economic consequences. Issues in Science and Technology 2(1):125-133, for commentary on consequences of BST adoption.

Kalter, R. 1985b. Testimony before the House Subcommittee on Investigations and Oversight of the Science and Technology Committee. April 17.

Kalter, R. J., and McGrath, W. 1985. Biotechnology. In New York Agriculture 2000. Office of Governor, State of New York, Albany.

Kalter, R. J., Milligan, R. , Lesser, W., McGrath, W., and Bauman, Dale. 1985. Biotechnology and the dairy industry: production costs and commercial potential of the bovine growth hormone. Agricultural Economics Research 85-20. Ithaca: Cornell University.

Kloppenburg, J. Jr. 1984. The social impacts of biogenetic technology in agriculture: past and future. In The social consequences and challenges of new agricultural technologies, eds. G. Berardi and C. Geisler, pp. 291-321 . Boulder: Westview.

Larson, O. 1947. Ten years of rural rehabilitation in the United States. Washington, D.C.: Bureau of Agricultural Economics, USDA.

Lipton, M. with Longhurst, R. 1985. Modern varieties, International agricultural research and the poor. Washington, D. C.: World Bank, CGIAR.

McAllister, J. M. 1981. In Extension education and rural development, vol. I , eds. Crouch and Chamola. New York: John Wiley, p. 135-145.

MacFadyen, J. T. 1984. Gaining ground. New York: Ballantine.

McWilliams, R. I. 1971. Social aspects of farm mechanization in Oklahoma. Bulletin No. B-339. Stillwater, Oklahoma: Agricultural Experiment Station.

Morris, J. 1983. Economics of small farm mechanization. In Small farm mechanization for developing countries, eds. P. Crossley and J. Kilgour, pp. 171-184, New York: John Wiley.

National Agricultural Research and Extension Users Advisory Board. 1987. Appraisal of the proposed 1988 budget for Food and Agricultural Sciences. Report to the President and Congress. Washington, D.C.: USDA.

New York State Senate. 1985. Farmers in transition: when hard work was not enough. Albany.

Office of Technology Assessment. 1986. Technology, public policy, and the changing structure of American agriculture. Office of Technology Assessment, U.S. Congress, Washington, D.C., p. 53.

Perrin, R., and Winkelmann, D. 1976. Impediments to technical progress on small versus large farms. American Journal of Agricultural Economics 58(5):888-894.

Pimentel, D. 1987. Down on the farm: genetic engineering meets ecology. Technology Review 5:24-30.

Prahladachar, M. 1983. Income distribution effects of the green revolution in India: a review of empirical evidence. World Development 11:929-30.

Rogers, E. M. 1983. Diffusion of innovations. (3rd Ed.). New York: Free Press

Rogers, E. M. 1962. Diffusion of innovations. New York: Free Press.

Rogers, E. M., and Kincaid, D. L. 1981. Communication networks: toward a new paradigm for Research. New York: Free Press.

Rogers, E. M., and Shoemaker, F. F. 1971. Communications of innovations: a cross cultural approach. New York: Free Press.

Rogers, E. M., R. J. Burdge, P. F. Korsching, and J. F. Donnermeyer. 1988. Social change in rural societies: an introduction to rural sociology. Englewood Cliffs, N.J.: Prentice Hall (Third edition).

Roling, N., Ascroft, J. R., and Chege, F. W. A. 1976. The diffusion of innovations and the issue of equity in rural development. Communication Research 3(2):155-170.

Ruthenberg, Hans. 1985. In Innovation policy for small farmers in the tropics. ed. by H. E. Jahuke. Oxford: Clarendon Press.

Ruttan, V. W. 1986. Increasing productivity and efficiency in agriculture. Science 231:1.

Ruttan, V. W. 1982. Changing role of public and private sectors in agricultural research. Science 216 :23-29.

Ruttan, V. W. 1977. The green revolution: Seven generalizations. International Development Review 19(4):16-23.

Saint, W. S., and Coward, E. W., Jr. 1977. Agricultural and behavioral sciences: emerging orientations. Science 197:733-737.

Savill, R. J. 1941. Trends in mechanization and tenure changes in the southeast. In The people, the land and the church in the rural south. Chicago: The Farm Foundation.

Schuh, G. E. 1987. The role of agricultural research in the world economy. Presentation at the Hatch Centennial Symposium, Cornell University, Ithaca, N. Y., May 4-5.

Schulman, M. 1987. Bovine growth hormone: who wins: who loses: what's at stake? The new farm, Part 1 in 9(6):36, 38-39; Part 2 in 9(7):28-41.

Schmitz, A., and Seckler, C. 1970. Mechanized agriculture and social welfare: the case of the tomato harvester. _American Journal of Agricultural Economics_ 54 569-77.

Shaner, W. W., Phillip, P. F., and Schmehl, W. R. 1982. _Farming systems research and development: guidelines for developing countries_. Boulder: Westview.

Singer, H. 1982. _Technology for basic needs_. Geneva: International Labor Office.

Small farm equipment for developing countries. 1985. Proceedings of the International Conference on Small Farm Equipment for Developing Countries: Past Experience and Future Priorities. Manila: The International Rice Research Institute, September.

Street, J. H. 1957. _New revolution in the cotton economy: mechanization and its consequences_. Durham: U. of No. Carolina Press.

Tangley, L. 1986. Biotechnology on the farm. _Bioscience_ 36(9):590-594.

Taylor, P. S. 1938. Power farming and labor displacement in the cotton belt, 1937. _Monthly Labor Review_ 46(March):595-607.

Williams, S. and Karenoffers, R. 1986. _Agribusiness and the small-scale farmer_. Boulder, Colo.: Westview.

Wolf, E. C. 1986. Beyond the green revolution: new approaches to Third World agriculture. World Watch Paper No. 73, Washington D.C.

World Bank. 1975. Land reform. _World Bank sector policy paper_. Washington, D.C. (May).

Wright, A. 1984. Innocents aboard: American agricultural research in Mexico. In _Meeting the expectations of the land_, eds. W. Jackson, W. Berry, and B. Colman, pp. 135-51, San Francisco: North Point Press.

Young, A. L. 1987. Agricultural and federal science policy. Address to the Hatch Centennial Symposium, Cornell University, May 4-5, Ithaca, N. Y.

12. Conclusion:
Biotechnology, Farming, and
Agriculture into the 21st Century

Biotechnology offers the prospect of a vigorous and responsive food and fiber system for the United States and the world in general. We have already seen a decline in longstanding barriers to yield, vulnerability to pests, and limitations on product quality (Hess et al. 1987). This volume has outlined some of the ways new research techniques and technological developments are affecting the course of agricultural science, the availability of new products, and the implementation of these developments by farm operators.

The specific impacts or consequences of biotechnology in agriculture can be overdrawn. Progress continues to be made by conventional means of scientific inquiry that are inseparable from the novel techniques. Nevertheless, biotechnology is affecting the agricultural research system, the network of firms and corporations that sells inputs to farmers, as well the conduct of farming itself. These changes are occurring along identifiable paths (Yoxen 1986). This chapter attempts to provide an overview of the major themes or directions of change associated with the advent of biotechnology in agriculture.

CENTRAL DIMENSIONS OF CHANGE IN U.S. AGRICULTURE

Institutional Alterations

Although the reported share of expenditures allocated to biotechnology research continues to grow, it is not clear whether such a category can be clearly separated from other approaches based on more traditional techniques (Abelson 1984). If anything, the concept of biotechnology has given agricultural science a thematic perspective on the future of food and fiber production at a time when agricultural research was becoming moribund (National Academy of Sciences 1972; 1982; Rockefeller Foundation 1982).

It is clear that the rate at which the potential of biotechnology is being realized through new products for the farmer, processor, and consumer is increasing rapidly. Formerly, the changes wrought by biotechnology were largely confined to the institutions that do agricultural research, the regulatory apparatus that approves new products and writes guidelines for the conduct of federally-sponsored research, and the corporate sector where chemical and agribusiness firms are drawn by the prospect of great profits (Kloppenburg 1988).

The changes associated with biotechnology in agriculture emanate from the laboratory in increasingly broader and diffuse waves. New techniques for manipulating and altering the performance qualities of plants, microbes, and animals have affected agricultural research, extension, and teaching institutions. Researchers can continue to practice normal science in agriculture using conventional scientific techniques, but increasingly the frontiers of knowledge in the various disciplines are incorporating fundamental advances in molecular biology to achieve applied research objectives in more rapid and effective ways (Thurow 1987b; Buttel et al. 1984). The new approaches also give rise to new possibilities not previously attainable with conventional techniques.

In this way, agricultural departments, disciplines, and training programs have been forever changed by the advent of biotechnology. Larger institutional collectivities in turn are affected by shifting priorities for research equipment and the human resources to go with it. Competition for well-trained senior practitioners of biotechnology research techniques raised salary levels and attracted extramural support. Thus, it might be said that biotechnology widened the distance or inequality among institutions in terms of the level of science practiced, as well as in shares of the monetary base available for agricultural research (Buttel 1985).

The increasingly specific division of labor among agricultural universities often precipitates internal reorganizations or redirection of effort in light of the potentials of biotechnology. Criticisms of existing directions in light of the new possibilities of biotechnology led to restructuring of the USDA Agricultural Research Service to better respond to the emerging research environment. Many universities launched biotechnology research centers or programs, often with state-appropriated funds for public universities. Several prestigious private universities also launched biotechnology research centers in partnership with large U.S. and overseas corporations (Buttel 1985; 1986).

At the land grant universities, criticisms from the Pound report and the Winrock report caused a redirection and realignment in the agricultural experiment station system. Limited programs of competitive research funding were instituted, while formula-funding of agricultural research was allowed only nominal increases in funding (GAO 1986). More recently the role of adaptive research by publicly-funded agricultural research institutions has been recognized as a necessary and useful balance to basic research having less direct consequences for agricultural producers. Nevertheless, the techniques and approaches of biotechnology are becoming routinized tools for many plant and animal scientists because, more than ever before, they permit research objectives to be realized in more practical and efficient ways (Thurow 1987a).

Industrial Reorganization

The balance of basic research in the plant and animal sciences seems to have shifted toward the private sector, as measured by dollar volume of expenditures. Developments in the legal-regulatory arena have provided clear incentives for private firms to invest in research and product development for agricultural producers. Patent protection for new plant varieties, coupled with new capabilities for modifying and installing performance characteristics, have fueled major investments by seed and petrochemical firms (Yoxen 1984).

A major shift in the structure of the seed and

other input industries has accompanied the rise of
biotechnology. Larger companies have absorbed smaller
ones to increase market share and consolidate domina-
tion of specific markets. Correspondingly, the larger
agribusiness firms often are able to devote greater
resources to product development, approval, and marketing.

The entreprenurial potentials of biotechnology
have not been lost on individual scientists. Small
private companies that are able to identify and develop
a marketable product offer the prospect of great wealth
for the inventor and the firm's investors if the firm
is acquired by a larger company or if a product
achieves commercial success (Fairtlough 1986). As
participants in small startup companies, or in
consultative relationships with large corporations,
university professors have played a pivotal role in
establishing commercial applications for genetic
engineering (Kenney 1986, p. 90).

The dominance of larger firms in a restructured
and concentrated agribusiness sector is increased by
the potential for sudden shifts in input markets associated with product substitution derived from biotechnology. New products such as plants that exude their
own pesticides, fix nitrogen, or have an installed
resistance to a particular herbicide can induce major
displacements of chemicals for insect, weed, or fungus
control, as well as reducing the need for purchased
nutrients. Firms with specialized product lines tied
to particular agricultural industries are especially
vulnerable to such changes. Larger organizations of-
fering diversified products to multiple agricultural
industries have a competitive edge under such conditions, shaping an industrial structure concentrated in
fewer, larger firms (Prentis 1984).

In an era of rapidly declining farm numbers and
concerns about the cost, quality, and reliability of
the American food system, biotechnology is alternatively viewed as both a threat and salvation. A major
concern associated with potent new biotechnology
products is the industrial reorganization and
concentration of market power that they often bring
(Doyle 1985). The division of profits, benefits, and
costs among large firms, small firms, farmers, and
consumers has been considered from the vantage points
of fairness, social justice, and corporate responsibility.

Another aspect favoring larger agribusiness firms is the costly and lengthy regulatory process associated with new items to be introduced into the environment (Brill 1985). The process is particularly difficult for biotechnology products because of the unknown risks of biotechnology (Slovic 1987; Perrow 1984; Miller 1983). The politics of regulation are such that many incentives to the regulator lie in caution and delay (Jaffe 1987; Teich et al. 1985). The costs of a potentially unsafe or damaging product are difficult to balance against the approval-seeking firm's sunk costs of research and development, or its opportunity costs associated with not being allowed to market the new item.

An unsettled regulatory environment is not advantageous to the small firm delayed in recovering its investment with few or no other sources of income (Office of Technology Assessment 1984). Furthermore, uncertainty in the regulatory process is often translated into expanded demands for data attesting to the safety, efficacy, and other properties of the new product, which is another fixed cost more burdensome to the small firm.

Structural Change in Farming

Technical change has always been a source of structural change in agricultural industries (Office of Technology Assessment 1986). Biotechnology offers the prospect of more rapid shifts in industry structure associated with wide differences in advantage between adopters and non-adopters of biotechnology products. Ostensibly BST in the dairy industry is an example of such a paradigm-shifting innovation, yet other developments in other industries may exert even more pervasive effects on the operation, profitability, and survival of individual firms.

The situation of farm operators relative to their technology-adopting competitors has been described as an agricultural treadmill (Cochrane 1979). The analogy is useful, although biotechnology products are only beginning to reach the farm, although products are imminent in a number of agricultural industries. The wave of innovations that is coming will undoubtedly speed up the treadmill, further distancing users from nonusers of the new products.

The biotechnology origins will be transparent to the farmer for some items, readily substituting for the old seed or substance previously used in the production process. Other biotechnology products will require enhanced management and technical understanding to realize full benefits from the innovations. The latter genre of advances will have the greatest potential to differentiate the structure of the various agricultural industries. Nevertheless, when the innovations become available, adoption is neither instantaneous nor universal, and the rate of change may not be as rapid as expected.

Biotechnology is dawning on a world agricultural system that increasingly pits farmers in different nations against one another for market share, as exports are the lifeblood of most agricultural industries in the developed nations (Crott 1986). Nation-states and the politically-effective farmers they protect face major dilemmas in the coming decades. As long as the supply of food and fiber grows faster than world demand for the commodities, prices will remain low (Rees 1985).

The least-profitable farms and farm-industries in nation-states will be forced to absorb losses or lower profits. Farms can shift to other enterprises or go out of business. Nation-states could allow farms to decline in numbers or accept the reality that some foreign suppliers are more capable of producing the commodity. Down-sizing agriculture to a set of larger, more competitive firms that do not receive subsidies is one policy option. This is not what has occurred in practice.

Farming and agriculture are intertwined with national identify in multiple and diverse ways. Farmers are a vocal minority that have achieved a great deal of protection from market pressures through price supports, import quotas, and other protectionist interventions. Biotechnology lowers production costs and increases yields, making the costs of oversupply and farm subsidy even greater (Reilly 1988).

Nation-states that intervene in agricultural markets face difficult political and financial decisions about the level of support they are willing to extend their nation's farmers. As supply increases and prices fall below target levels, payments to farmers become a growing political issue. The specific features of these scenarios will be highly variable across nation-states, yet the dairy industry is becoming an increasingly costly ward of the state under present technological conditions. When biotechnology products appear, excess supply and embarrassing levels of subsidy may become salient public policy concerns (Fallert et al. 1987).

Biotechnology is not inherently detrimental to family farming. It will only accelerate the process by which the United States and other developed nations face fiscal choices about symbols of national heritage and the realities of resource endowments and comparative advantage in the world system. Family farming has certain managerial and organizational advantages over corporate and non-owner operation of farm firms (Molnar 1986). Whether biotechnology developments will force family farms toward industrial-scale levels of operation remains an empirical question. Clearly, the situation will evolve in a context of world competition, government intervention, and agribusiness efforts to develop and market the innovations. In the United States, public agricultural research will continue to play an adaptative role for the states and regions as well as serving as a defense network against pests and disease, and in identifying acute and cumulative impacts of agricultural activity on the environment.

Environmental Impacts

A major set of issues that will continue to complicate the implementation of biotechnology in agriculture pertains to the unknown and perhaps indeterminable effects of altered lifeforms on existing species and ecosystems. The formidable barriers of regulation and public concern have shifted some of the biotechnology research strategy toward identification of superior existing organisms that then can be propagated and marketed through traditional channels. Even if naturally superior microorganisms are identified, the transfer of biotic forms from place to place offers unknown risks associated with species displacement, disease, or parasites. The survival and reproduction of altered microorganisms may be an unlikely event, but the risks have become public issues for which no absolute assurances can be truthfully offered.

Environmental themes are dominating the policy agenda in agriculture. Groundwater pollution, pesticide safety, and the disposal of animal waste present multifaceted dilemmas for many agricultural industries and production regions. The presently available solutions are costly and may prohibit some types of production in some locales.

Biotechnology offers the prospects for achieving a sustainable agricultural program in the United States and the world. If plants can be induced to exude their own pesticides, if animals can be designed to produce more healthful meat, and if organisms can be developed that more effectively fix atmospheric nitrogen and digest animal waste, biotechnology will have realized the promise of even its most fervent exponents. Thus, environmental matters represent both the greatest barriers to the utilization of biotechnology in agriculture, as well as the greatest prospects for contributions to society.

CONCLUSION

Biotechnology is not likely to be the bane or boon its detractors and promoters make it out to be. With some degree of certainty it can be asserted that biotechnology is a major factor driving the evolution of agricultural production systems in the world today.

Biotechnology, however, is only one of an array of shifting technological frontiers altering the practice of farming around the world (Batie and Healy 1980). Advances in computing and information processing increase the quantity, timeliness, and accuracy of information available to the producer, input supplier, and processor. Improvements in irrigation, machinery, and material handling, not to mention robots, have many yet-to-be realized consequences for the food and fiber system. Corollary developments in agricultural markets use new financial instruments to shift risk and increase the pricing and operational efficiency of the food marketing system. Thus, biotechnology is only one of many sources of change in agriculture, not considering the marked effects national policy and international competition can have on farming and agribusiness.

Biotechnology may be the most intriguing front of

advance in agriculture because it has to do with the manipulation and change of living things, an arena not heretofore open to direct specification and control. Perhaps it is just this lack of precedence and proximity to the fundamental mysteries of life and the world around us that has driven the recent evolution of our agricultural institutions. Similarly, the concept may be equally potent in motivating farmers and their sympathizers to be alarmed about adjustments in an industry and way of life that views itself as fundamental to national character and identity. Biotechnology seems to offer the prospect of shifting the traditionally rural, family, and locality-based farm industry toward increased external influence and control.

Safety and environmental issues aside, many populist explanations for the recent difficulties experienced by farming and agriculture locate much of the source of the problem in government involvement in agriculture and the actions of large corporations that are thought to exploit the farmer. Biotechnology can be viewed as epitomizing the unwanted future in terms of the disagreeable recent past. Regardless of the specific manifestations of biotechnology and the way it is enacted in specific agricultural industries, it will remain symptomatic of larger trends that seem to be inexorably changing the face of agriculture in this country and the world. Reconciling values about farming, democracy, and our economic system with the potentials of new technology and the imperatives of the world system will occupy agricultural and national leaders for some time to come.

REFERENCES

Abelson, P., ed. 1984. Biotechnology and biological frontiers. Washington, DC: The American Association for the Advancement of Science.

Batie, S. S. and Healy, R. G., eds. 1980. The future of American agriculture as strategic resource. Washington, DC: The Conservation Foundation.

Brill, W. J. 1985. Regulation of biotechnology, Science 227:381-84.

Buttel, F. H. 1985. The land-grant system: a sociological perspective on value conflicts and ethical issues. Agriculture and Human Values 6:78-95.

Buttel, F. H. 1986. Biotechnology and agricultural research policy: emergent issues. In _New directions for agriculture and agricultural research_, ed. K. A. Dahlberg. Totowa, NJ: Rowman and Allenheld.

Buttel, F. H.; Cowan, J. T.; Kenney, M.; and Kloppenburg, J., Jr. 1984. The political economy of agribusiness reorganization and industry-university relationships. _Research in Rural Sociology and Development_ 1:315-43.

Carson, R. 1962. _Silent spring_. Boston: Houghton-Mifflin.

Cochrane, W. W. 1979. _The development of American agriculture_. University of Minnesota Press: Minnesota.

Crott, R. 1986. The impact of isoglucose on the international sugar market. In _The biological challenge_, eds. S. Jacobsson, A. Jamison and H. Rothman. Cambridge: Great Britain at the University Press, pp. 96-123.

Doyle, Jack. 1985. _Altered harvest_. New York: Viking Penguin.

Fallert R.; McCuckin, T.; Betts, C.; and Brunner, G. 1987. _BST and the dairy industry_. Washington, DC: USDA-ERS-AER report 570.

Fairtlough, G. H. 1986. Genetic engineering-problems and opportunities. In _The biological challenge_, eds. S. Jacobsson, A. Jamison and H. Rothman. Cambridge: Great Britain at the University Press, pp. 12-36.

General Accounting Office. 1986. Biotechnology: analysis of federally-funded research. RCED-86-187. Washington, DC: U.S. General Accounting Office.

Hess, C. E. et al. 1987. _Agricultural biotechnology_. Washington, DC: National Academy Press.

Jaffe, G. A. 1987. Inadequacies in the federal regulation of biotechnology. _Harvard Environmental Law Review_ 11:491-550.

Kenney, M. 1986. _Biotechnology: the university-industrial complex_. New Haven and London: Yale University Press.

Kloppenburg, J. 1988. _First the seed; the political economy of plant biotechnology_. New York: Cambridge University Press.

Miller, J. D. 1983. Scientific literacy: a conceptual and empirical review. _Daedalus_ 112:29-48.

Molnar, J. 1986. _Agricultural change: consequences for southern farms and rural communities_. Boulder: Westview Press.

National Academy of Sciences. 1982. Genetic engineering of plants. Washington, D.C.: National Academy of
Science.

Office of Technology Assessment. 1984. Commercial
biotechnology: an international analysis. Washington, D.C.: U.S. Congress, Office of Technology Assessment.

_____. 1986. Technology, public policy and
the changing structure of agricul ture. Washington,
DC: U.S. Congress.

Perrow, C. 1984. Normal accidents: living with
high-risk technologies. New York: Basic Books,
Inc.

Prentis, Steve. 1984. Biotechnology: a new industrial revolution. New York: George Brazilier.

Reilly, J. 1988. Cost-reducing and output-enhancing
technologies. Technical Bulletin 1740. Washington:
USDA-Economic Research Service.

Rees, J. 1985. Natural resources: allocation, economics and policy. London and New York: Methuen.

Slovic, P. 1987. Perceptions of risk. Science
236:280-85.

Teich, A. H.; Levin, M.A.; and Pace, J. H. eds.
1985. Biotechnology and the environment. Washington, DC: American Association for the Advancement of
Science.

Thurow, L. 1987a. A weakness in process technology.
Science 238:1659-63.

_____. 1987b. Agricultural institutions and
arrangements under fire. Mimeo. Cambridge: Massa-
chusetts Institute of Technology, Sloan School of
Management, Department of Economics.

Yoxen, E. 1986. The social impact of biotechnology. Trends in Biotechnology 4:86-88.

_____. 1984. The gene business. New York: Har
per and Row.

About the Contributors

Joseph J. Molnar is professor of rural sociology at Auburn University. He has presented numerous papers and written many articles on organizational dynamics, rural community development, agricultural economics, and environmental sociology. Most recently, he edited the volume <u>Agricultural Change: Consequences for Southern Farms and Rural Communities</u> (Westview Press, Boulder, CO, 1986). He has served in various administrative capacities and has chaired several academic research and technical committees. He is a member of a number of sociological and agricultural societies and honoraries. He received a Ph.D. degree in sociology from Iowa State University and M.A. and B.A. degrees from Kent State University.

Henry Kinnucan is assistant professor of agricultural economics at Auburn University. He has published numerous articles on price policy, industrial organization, taxation, and marketing. He is active on the regional technical committee "Economics of Biotechnology in Agriculture" and others. He belongs to international, national, and regional economics associations, and he has received professional awards for excellence in research discovery and contribution. He received Ph.D. and M.S. degrees from the University of Minnesota and a B.S. degree from the University of Illinois.

N.C. **Brady** is Senior Assistant Administrator for Science and Technology at the U.S. Agency for International Development, U.S. Department of State. He also serves as editor of <u>Advances in Agronomy</u> and he is Professor of Soil Science Emeritus at Cornell University. He has held various academic and government posts,served on national panels and boards, fulfilled international consultancy assignments, and been active in professional organizations. He received a Ph.D. in soil science from North Carolina State University and a B.S. in chemistry from Brigham Young University, and he has been awarded a number of academic honors.

Lawrence Busch is professor of sociology at the University of Kentucky. His field of specialization is agricultural research organization and policy. He has coedited <u>Science, Agriculture, and the Politics of Research</u>, with William B. Lacy (Westview Press, 1983) and the forthcoming <u>Second Nature; Plant Breeding and the New Biotechnologies</u>, with William B. Lacy, J. Burkhardt, and M. Hansen (Blackwell, London). He has also written numerous articles on agricultural research and research policy and ethics, and he is active in several professional associations. He received Ph.D. and M.S. degrees from Cornell University and a B.S. degree from Hofstra University.

Frederick H. Buttel is professor of rural sociology at Cornell University. His areas of expertise are rural, political, and environmental sociology, and technology and social change. He has coedited or cowritten several volumes, including <u>The Rural Sociology of the Advanced Societies</u> (Allanheld, Osmun, 1980). He has held office for professional groups and been elected Fellow of the AAAS. He received Ph.D. and M.S. degrees from the University of Wisconsin and an additional M.S. from Yale University.

E. Melanie DuPuis is a Ph.D. student in rural sociology at Cornell University. Her dissertation research focuses on a social history of dairy technology in New York State. She is an author of a recent publication in <u>Bio Science</u>. She received a M.S. degree in rural sociology at Cornell University and a B.A. from Radcliffe College.

Robert E. Evenson is professor of economics at Yale University. He specializes in the economics of technology and of agricultural development. He has held several academic posts in economics and agricultural economics and has been visiting lecturer and professor overseas. He has edited, coedited, and cowritten several books, including <u>Agricultural Research and Productivity</u>, with Y. Kislev (Yale University Press, New Haven, CT, 1975). He has also written numerous papers and articles on agrarian technology and research, and on agricultural economic and labor issues. He received from the University of Chicago a Ph.D. in economics, and from the University of Minnesota a M.S. in agricultural economics and a B.A. with Highest Distinction in agricultural business administration.

Charles C. Geisler is associate professor of rural sociology at Cornell University. His particular areas of interest are natural resource sociology, social impact assessment, and agriculture and land reform. He has coedited <u>The Social Consequences and Challenges of New Agricultural Technologies</u>, with G. Berardi (Westview Press, 1984), and he has written a background paper for the Office of Technology Assessment on equity and sustainability in agriculture. He serves on the editorial boards of several journals and belongs to various professional societies. He received Ph.D. and M.A. degrees from the University of Wisconsin, Madison, and a B.A. degree from Dartmouth College.

Upton Hatch is assistant professor of agricultural economics at Auburn University. His specific field is technical change and resource use. He has contributed to a volume, <u>Biotechnology in Agricultural Chemistry</u>, edited by Homer M. LeBaron (American Chemical Society, Washington, D.C., 1987), and he has written articles and lectured on the economic impact of bovine growth hormone. He is a member of several agricultural economics and resource organizations. He received a Ph.D. degree from the University of Minnesota, a M.S. from the University of Georgia, and a B.A. from Dartmouth College.

Martin Kenney is assistant professor of agricultural economics and rural sociology at the Ohio State University. He specializes in university-industry relations in biotechnology, the impacts of biotechnology on the Third World, and the role of venture capital in economic development. He has published the volume <u>Biotechnology: The University-Industrial Complex</u> (Yale, 1986), as well as journal articles on biotechnology and the Third World and on the venture capital industry. He received his Ph.D. degree from Cornell University.

Ronald D. Knutson is professor and extension economist in the Department of Agricultural Economics at Texas A&M University. His areas of expertise are agricultural policy and dairy marketing. He has published the volume <u>Agricultural and Food Policy</u> (Prentice Hall, Englewood Cliffs, NJ, 1983), and he has cowritten articles and conducted studies on technological change and the structure of agriculture. He received Ph.D. and B.S. degrees from the University of Minnesota and a M.S. degree from the Pennsylvania State University.

Fred Kuchler is an economist with the Resources and Technology Division of the Economic Research Service, U.S. Department of Agriculture. His focus of investigation is industrial organization and public choice. He has written articles about <u>ex ante</u> technology assessment, economic impacts of regulating agricultural technology (pesticides and genetic engineering), and estimating the value of programs to control exotic crop pests. He has twice received the Administrator's Special Merit Award for Outstanding Research (Economic Research Service). He received Ph.D. and M.A. degrees in economics from Virginia Polytechnic Institute and a A.B. in economics from the University of California, Davis.

William B. Lacy is professor and director of the Food, Environment, and Agriculture Program in the Department of Sociology at the University of Kentucky. His special field of interest is organizational behavior and the sociology and social psychology of science and agriculture. He has cowritten and coedited several books, including <u>Food Security in the United States</u> (Westview Press, 1984), with Lawrence Busch. He has received grants and contracts from, and served as consultant for, several government agencies and national foundations. He received Ph.D. and M.A. degrees from the University of Michigan, another M.A. from Colgate University, and a B.S. degree from Cornell University.

James W. Richardson is professor of agricultural economics at Texas A&M University. He specializes in agricultural policy and technology evaluation. He has written numerous articles and bulletins on the effects of changes in policy and technology on American farmers. His work emphasizes the impacts of policy on representative crop and livestock farms. He is a member of several agricultural economics associations and has been honored with awards for quality of communication and for an outstanding extension education program. He received Ph.D. and M.S. degrees from Oklahoma State University and a B.S. degree from New Mexico State University.

Clair E. Terrill is retired from the position of collaborator with the Agricultural Research Service, U.S. Department of Agriculture. His area of expertise is the genetics and reproduction of sheep and goats. He has written articles on the improvement, breeding, reproduction, and production of sheep and goats. He has received a number of professional honors, including "All-Time Great in the World of Animals" from the Agriservices Foundation and induction into the International Stockmen's Hall of Fame, as well as several industry service awards and election to the presidencies of the American Genetic Association and the American Society of Animal Science. He received a Ph.D. degree from the University of Missouri and a B.S. degree from Iowa State University.

Robert D. Yonkers is a research associate and Ph.D. candidate in the Department of Agricultural Economics at Texas A&M University. His focus of interest is agricultural policy and marketing systems. He belongs to several agricultural associations. He received a Master's degree in agriculture from Texas A&M University, and he earned a B.S. degree from Kansas State University.

Author Index

Subject Index

Printed and bound by CPI Group (UK) Ltd, Croydon, CR0 4YY

23/10/2024

01778240-0016